Laboratory Lifestyles

Leonardo

Roger F. Malina, Executive Editor

Sean Cubitt, Editor-in-Chief

See http://mitpress.mit.edu for a complete list of titles in this series.

Laboratory Lifestyles

The Construction of Scientific Fictions

edited by Sandra Kaji-O'Grady, Chris L. Smith, and Russell Hughes

The MIT Press
Cambridge, Massachusetts
London, England

This book was set in ITC Stone Serif Std by Toppan Best-set Premedia Limited. Printed and bound in the United States of America.

Library of Congress Cataloging-in-Publication Data

Names: Kaji-O'Grady, Sandra, editor.
Title: Laboratory lifestyles : the construction of scientific fictions /
   edited by Sandra Kaji-O'Grady, Chris L. Smith, and Russell Hughes
Description: Cambridge, MA : The MIT Press, 2018. | Includes bibliographical
   references and index.
Identifiers: LCCN 2018011213 | ISBN 9780262038928 (hardcover : alk. paper)
Subjects: LCSH: Laboratories. | Work environment. | Architecture--Human
   factors. | Architecture and society.
Classification: LCC Q183 .L27 2018 | DDC 001.4--dc23 LC record available at https://lccn.loc.
gov/2018011213

10 9 8 7 6 5 4 3 2 1

Contents

10   The Urbane Laboratory: Applied Sciences New York   219
Russell Hughes

Series Foreword

Leonardo/International Society for the Arts, Sciences, and Technology (ISAST)

Leonardo, the International Society for the Arts, Sciences, and Technology, and the affiliated French organization Association Leonardo, have some very simple goals:

1. To advocate, document, and make known the work of artists, researchers, and scholars developing the new ways in which the contemporary arts interact with science, technology, and society.
2. To create a forum and meeting places where artists, scientists, and engineers can meet, exchange ideas, and, when appropriate, collaborate.
3. To contribute, through the interaction of the arts and sciences, to the creation of the new culture that will be needed to transition to a sustainable planetary society.

When the journal *Leonardo* was started some fifty years ago, these creative disciplines usually existed in segregated institutional and social networks, a situation dramatized at that time by the "Two Cultures" debates initiated by C. P. Snow. Today we live in a different time of cross-disciplinary ferment, collaboration, and intellectual confrontation enabled by new hybrid organizations, new funding sponsors, and the shared tools of computers and the Internet. Sometimes captured in the "STEM to STEAM" movement, new forms of collaboration seem to integrate the arts, humanities, and design with science and engineering practices. Above all, new generations of artist-researchers and researcher-artists are now at work individually and collaboratively bridging the art, science, and technology disciplines. For some of the hard problems in our society, we have no choice but to find new ways to couple the arts and sciences. Perhaps in our lifetime we will see the emergence of "new Leonardos," hybrid creative individuals or teams that will not only develop a meaningful art for our times but also drive new agendas in science and stimulate technological innovation that addresses today's human needs.

For more information on the activities of the Leonardo organizations and networks, please visit our Web sites at http://www.leonardo.info/ and http://www.olats.org/. Leonardo Books and journals are also available on our ARTECA art science technology aggregator: http://arteca.mit.edu/.

Acknowledgments

This book is one of the outcomes of a research project conducted by Kaji-O'Grady and Smith with funding from the Australian Research Council (ARC) through a Discovery Grant (2013–2016), titled "From Alchemist's Den to Science City: Architecture and the Expression of Experimental Science." We are deeply grateful to the anonymous reviewers who saw value in our original proposal. Additional support in the production of the book was granted from the Faculty of Engineering, Architecture and Information Technology at the University of Queensland.

While the ARC's funding made it possible for the editors to recognize the emergence of lifestyle science and to frame and curate the project, this book represents the efforts of a further twelve authors drawn from science, medicine, the history of science, technology studies, anthropology, political science, art, and architecture. Each author has been supported by their home institutions; by architects, scientists, and building managers who facilitated visits to individual buildings and related their views and experiences; by librarians who permitted access to archives; and by colleagues who read and commented on their work. The editors would like to thank Doug Sery and Sean Cubitt who saw value in the project.

Figures

Preface

This book responds to the phenomenal investment in new buildings for science in the last decade. Beyond a handful of architecturally ambitious laboratories constructed by elite institutions in the second half of the twentieth century, the laboratory building type was widely considered unglamorous and pragmatic. Those involved in publicizing individual new laboratory buildings—from journalists to scientists themselves—continue to celebrate well-designed laboratory buildings as exceptional departures from the type, yet this is no longer the case. The fate of the laboratory changed in 1980 when the nascent biotechnology company Genentech went public, raising USD 35 million in a matter of hours. Since then, the funding of basic research has increasingly engaged stock markets and venture capital. In turn, this has fueled an appetite for spectacular architecture for laboratory facilities with luxurious amenities on verdant campuses or prominent urban sites. This new generation of laboratory buildings are designed to lure the necessary venture capital, philanthropy, and star scientists, essential to their operation. Their architecture has sought to imbue centers of science with both novelty and "lifestyle" elements intended to induce creativity, enhance collegiality, foster ambition, and convey prestige, for the purpose of securing higher yields of scientific discovery and production.

It is the migration of luxurious architecture and what we refer to as "lifestyle science" to the profit-driven realm of multi-tenant speculative laboratory buildings, however, that prompted us to instigate this critical volume, for it signals a wholesale change. The sector is dominated by corporations established between 1979 and 2004 on the back of the biotech boom, most notable being Nexus Properties, Alexandra Real Estate Equities, and BioMed Realty Trust. Throughout the 1980s and 1990s these companies developed laboratory buildings for prospective tenants in suburban research campuses. Back then, the buildings were predicated solely on the provision of technical laboratory services and maximum return and were indistinguishable from the big box office and retail outlets around them. Yet today we see prestige developments such as The Alexandria

Center for Life Science, in New York City, which opened the first two of its planned three towers in 2011. It is marketed to tenants on the basis of its East River views and the epicurean delights savored in Riverpark and Riverpark Farm, on-site restaurants "curated" by the celebrity chef Tom Colicchio. The Alexandria Center also has a large conference and event facility, an acre of open green park, a riverfront esplanade, and a prominently located gymnasium. BioMed Realty's newest project, a USD 189 million life-sciences laboratory and office campus in San Diego's University Towne Center, not only includes a conference facility, restaurant, fitness center, and courtyard space for recreation, it also boasts an event lawn, bocce and ping pong courts, an herb garden, and a dining terrace.

Tailored to compete for tenants in a competitive market, speculative laboratory buildings are a barometer of the financial health of the science research ecosystem, and of how scientific organizations see themselves and their employees. Richly appointed laboratory architectures, and the lifestyles they symbolically and spatially foster, signal a shift in the role of the scientist and what constitutes scientific practice. Is scientific labor only that which takes place in the realm of pipettes and beakers, when driving thermal cyclers and scrutinizing tissue samples? Are scientists working when stretching at a yoga class, playing volleyball in the annual company tournament, chatting with neighboring teams in a café, or showing off their extravagant facilities (and the corporate success they convey) to the visiting pharmaceutical executive? All of these activities take place in, and are anticipated by, new laboratory buildings. The gymnasiums and restaurants are there because they are seen as critical components of scientific work. Managers encourage their use. While the material expression of lifestyle science looks much like its digital technology neighbors—indeed, deploys the very same foosball tables and exhortatory rhetoric—it is characterized by specific tensions and histories that make it a distinctive site for study.

The scientific method rests on the controlled, repeatable experiment. The laboratory is that neutral space that secures the experiment from all that might threaten the fidelity of the scientists' observations—vibration, contamination, heat, and so on. In this scenario, it is patently the inclusion of the scientist—an unreliable and inconstant human subject—that threatens the credibility of the experiment. Historically, the risks posed by this scientist-figure to the experiment have been mitigated. The scientist was to be configured as a trustworthy gentleman and, later, the professionalization and training of scientists and their assistants helped to secure the validity of any experiment. The idea that the individual scientist works in teams with others and is a mere cog in the institutional machinery of peer review and oversight is a further mitigation. The dynamic subjectivity of the scientist is a theme we will return to repeatedly, but for now, it is enough to note that the laboratory is a space that stages the encounter between the scientist and the experiment in isolation from the world. This isolation presents a unique challenge to the design of laboratories when the permeability of space that accompanies lifestyle aspirations comes up against the necessity of the containment function. Yet it is the intellectual conundrum of how one reconciles the scientific method with the conjectural claims of lifestyle architecture that draws the attention of the contributing authors in this book.

Philosophical speculation, the aesthetics of design, and the affective register of sociality are outside of the scientific method, yet have become essential to fulfilling its ambitions. Scientists themselves are recast as experimental subjects in workplaces designed with the aim of inducing specific behavioral responses. In today's laboratories, the scientist is at once the experimental subject of new forms of *biopolitics* and the instigator of experiments on other organisms, be they microbial, animal, or human. While this is made explicit in projects such as Biosphere 2, the focus of chapter 8, our argument is that scientists everywhere have implicitly become subjects in an experiment that is conducted, by scientific standards, in a shockingly sloppy fashion. For despite all the talk about the impacts of workplace design there is scarce empirical evidence. The effects of co-location upon individuals previously housed apart are conflated with the inclusion of "sticky corridors," "breakout spaces," and a jaunty stair. Media releases and marketing spiel announcing the successes of new laboratory buildings are cited as authoritative evidence, when the genre more accurately reflects architectural fantasies and their resonance in scientific markets. New buildings are credited for the recruitment of star scientists, regardless of the myriad factors, such as equipment, salaries, colleagues, and location, that attract employees. But it is exactly this blurring between the scientist as one who experiments and the scientist as a subject of experimentation that animates the design of the contemporary laboratory. The history, architectural expression, symptoms, and effects of this social experiment are under discussion in this book.

Yet there is an even bigger story. The entanglement of design and managerial techniques in the service of advanced capitalism shapes scientists with such subtlety that they feel and act like autonomous, self-determining agents. The consumption of lifestyle experiences and products has transformed into the production of styled lives and bodies. Scientists actively shape their behavior and bodies to meet expectations about their fitness for the new entrepreneurial workplace, while also producing discoveries that support the very notion of life as something malleable, exchangeable, extensible, and reducible to lines of code.

While life itself is at stake, there has been too little serious examination of the intersections between science, politics, economics, aesthetics, and the architecture of the laboratory. Which is not to say that there are not books on laboratory design and planning. The provision of advice to architects and their scientist clients on the architectural ingredients necessary for creative scientific research has become a profitable niche for consultants, and fueled a spate of texts that inform readers "how to design a laboratory." *Laboratory Lifestyles* has no desire to add to those volumes. While inherently suspicious of the instrumental conception of architecture, the book does not set out to expose science as somehow mistaken in its pursuit of architectural innovation. The empirical basis for architecture as a means to increase research creativity is slight, but the chapters in this book do not accuse architects of overstating their importance for professional gain. It is, rather, that the contribution architecture *actually* makes goes unremarked. This book seeks to understand and reveal the ways in which contemporary architecture productively negotiates the incongruities between the subjectivity of lifestyle and the purported objectivity of the scientific method. This volume coins the term "lifestyle science" to capture this unusual coupling of calculation with sensation, and exactitude with eccentricity. In this book, we are particularly interested in how the scientist is shaped by his or her engagement with laboratory architecture, and the many vocational leisure and lifestyle offerings that now attend and surround the bench space. The book seeks to understand contemporary laboratory architecture as evidence of change in what science today is, or rather how it is changing in cadence with the dynamic, indeed tectonic, shifts in the global economy, ecology, and technology systems. We contend that these buildings are as central to the making of modern science as the scientific talent they incubate.

In the opening chapter, we outline the economic forces driving the construction of laboratory buildings and explore the social and political changes that generate their resort-like styling. In doing so, we unpack the concept of "lifestyle," beginning with its introduction at the turn of the last century. That moment coincided with dramatic post-Darwinian arguments about life's mutability and limits, wherein questions of class and its expression became embroiled with evolutionary models. Lifestyle has continued to be theorized, and this framing chapter introduces some of the key figures and their arguments that recur throughout the book, from Thorstein Veblen and Pierre Bourdieu, through to Maurizio Lazzarato, Félix Guattari, Bruno Latour, Jacques Derrida, and Gilles Deleuze. Alongside them, we introduce three scientists whose public personae illustrate key theoretical points—James Watson, J. Craig Venter, and Nina Tandon. With more than fifty crucial years in age between Watson and Tandon, we think these leading scientists reveal something of the changing demands on the scientific researcher and the ways in which those demands are met.

The book approaches the laboratory from multiple angles, using diverse sources, and through the individual disciplinary expertise of each of the contributors. One

approach, rooted in anthropology and ethnography, is signaled in the book's titular
homage to the seminal text *Laboratory Life: The Construction of Scientific Facts* (1979),
by Bruno Latour and Steve Woolgar. Latour and Woolgar set out to observe the interactions between scientists at the bench as if they were an unknown tribe, with Latour
installed in the laboratory of Roger Guillemin at the Salk Institute for Biological Studies
between October 1975 and August 1977. Though Latour would commence his work
with a clear focus on the scientist, he would also come to explore the empirical objects
of science; those apparatuses or artifacts that have impact upon both science and the
scientist.[1] In chapter 2, Brian Lonsway and Kathleen Brandt argue that design "facts,"
and artifacts, are not only socially constructed, they also have a place in the construction of science. It is a point that resonates through the entire book, although Lonsway
and Brandt forensically pursue a single piece of evidence, the beanbags at the Xerox
PARC (Palo Alto Research Campus). This particular campus was built in 1970, just two
years after the introduction of the amorphous "Sacco," and its research environment
centered on this object that gains shape only in use. In revealing how the "lifestyling"
of one of information technology's (IT's) most critical research environments encouraged "disruption" within new labor management strategies, Lonsway and Brandt pro-
voke us to consider the effects of design artifacts in today's laboratories. In a similarly
forensic manner, in chapter 3, Albena Yaneva and Stelios Zavos turn to the National
Graphene Institute (NGI) opened in 2015. Yaneva is a former student of Latour's and
has since collaborated with him on a number of projects.[2] Yaneva and Zavos pursue
a study whose methodology is grounded in science and technology studies (STS) and
ethnography. They spent time with the scientists and architects of the NGI during the
development of the new building, observing their interactions and listening to their
conversations. Yaneva and Zavos find parallels and surprising differences between the
construction of design ideas and facts, and scientific ones.

Latour and Woolgar aspired to observing only the routine and normal practices of
science at the bench, avoiding the temptation of "instances of gossip and scandal" and
"sociological muckraking."[3] Jonas Salk applauded the exclusion. In his preface to *Laboratory Life*, Salk wrote: "The book is free from the kind of gossip, innuendo, and embar-
rassing stories, and of the psychologizing often seen in other studies or commentaries.
In this book, the authors demonstrate what they call the social construction of science
by the use of honest and valid examples of laboratory science."[4] It is an odd appeal,
given that Latour and Woolgar had concluded that the production and reception of scientific discoveries is distorted by its textual reliance and the social status of competing
individual scientists. It follows that Latour and Woolgar's own method and conclusions
are equally subjective, historic, and mediated by writing. In this book, we have been
keen to highlight different writing modes and authorial positions, to make clear the

ways in which textuality itself changes the view of the laboratory. We have not aspired to being passive or objective observers, or insisted our contributors avoid "sociological muckraking." Indeed, the question of what constitutes an "honest and valid" piece of evidence is not so quickly settled here. Several contributors stray far from Latour and Woolgar's declared devotion to that which is observed at the bench only. Steven Shapin neatly captures the thrust of the arguments made since the 1980s by those in STS, that science is "produced by people with bodies, situated in time, space, culture and society."[5] This broad construction of science is discussed and debated by our contributors who themselves come to the subject with unashamed prejudices, passions, and allegiances.

In chapter 4, *Laboratory Lifestyles* returns to Southern California where, historically, much of the ideological and practical content for lifestyle science germinated. In "The Beach Boys: Classified Research with a Southern California Vibe," Stuart Leslie details how some of America's smartest defense intellectuals preferred to "hang out" in some of its toniest beach communities: RAND in Santa Monica (1948), General Atomics in La Jolla (1964), and Hughes Research Laboratories in Malibu (1960). These places, designed to recruit, retain, and inspire scientists with scenic views, challenging problems, brilliant colleagues, and a lifestyle best described as "cold war avant-garde," perfected the art of concierge science and pioneered the geek chic that the rest of the world would soon follow. Leslie's contribution foregrounds Ksenia Tatarchenko's chapter 5, 'The Siberian Carnivalesque: Novosibirsk Science City." In Soviet Russia during the height of the Cold War period, the Novosibirsk Scientific Center or "Akademgorodok" became the epitome of Khrushchev's promise of affluence. Tatarchenko illuminates how the tension between the futuristic aspiration of the city's architectural forms and the archaic rituals of social interaction were enacted on these new utopian premises. Using the theories of Mikhail Bakhtin to understand how innovation is carnivalesque, Tatarchenko identifies Akademgorodok as a cybernetic descendant of the garden-city tradition or a Baconian "New Atlantis," or both. She explains how previous historians of science have mistakenly misrepresented Akademgorodok.

The middle section of the book examines key moments of consumption in the life of the laboratory. In chapter 6, "Scientific Dining," Sandra Kaji-O'Grady examines one aspect of laboratory life that affords management an opportunity to manipulate informal socialization and boost employee morale. The workday lunch is an occasion that combines informality with intimacy, business, and leisure. It is sandwiched by the concerns of the working day, yet is set outside of it as a "break." From the subsidized canteen of the 1980s to today's upmarket "foodie" restaurants, Kaji-O'Grady tracks the convergence of nutrition, commensality, and epicurean atmospheres that, quite literally, shape scientists. She notes that scientists have often complex relationships to food. Scientist Nina Tandon, for example, is a vegetarian, yet her research involves the use of animals as test subjects, demonstrating a particular discord. In chapter 7, Chris L. Smith explores this tension in regard to the consumption or "sacrifice" of animal lives deemed a necessary part of biological research. In chapter 7, "Naked in the Laboratory," Smith engages the posthumously published work of Derrida and explores the contemporary laboratory as an assemblage of human and nonhuman animals. He argues that the laboratory is an assemblage that spatially relegates value to different forms of life, attracting and rewarding some with sunny beachside lifestyles, and others with lives conceived and led entirely in artificially lit and ventilated basements. Through an exploration of the Parc de Recerca Biomèdica de Barcelona, opened in 2006, Smith considers the role of architecture in negotiating the different lives that come to constitute the laboratory, and finds a poignant moment in the subterranean animal facilities that sit adjacent to the complex's sporting and recreational facilities.

The lives and lifestyles that constitute a laboratory are a social and managerial experiment in which the scientist is the intended subject, but its instruments and effects are subtle—indeed, that is their point. Yet, there are moments where the inculcation of the scientist into the experimental scene is overt. In chapter 8, "Biosphere 2's Experimental Life," Tim Ivison, Julia Tcharfas, and Simon Sadler head to the middle of the Arizona desert where, in 1991, an independently financed, wildly ambitious, and blatantly performative ecology was built with human beings at the center of the experiment. Ivison, Tcharfas, and Sadler examine the notion of the Biosphere 2's "bionaut" as both scientist and subject whose research agenda explored connections between avant-garde theater, social psychology, space science, and the ecology movement. Biosphere is one of a family of futurist laboratories that included pneumatic structures, bio-domes, controlled environments, and utopian settlements in the earth's remotest and most inhospitable regions (including the deep ocean). In chapter 9, Nicole Sully, William Taylor, and Sean O'Halloran situate these "laboratories for living" in humankind's conquest of the globe and, from there, of outer space. Among President John F. Kennedy's goals was the establishment of a permanent American outpost and laboratory in space. As research toward the launching of interstellar laboratories continues apace, this chapter explores the intersection of science and science fiction, global politics, and experimental architecture. Russell Hughes in chapter 10, "The Urbane Laboratory: Applied Sciences New York," returns to earth, examining New York City's ostensible endeavor to become the new technology capital of the world. Congruent with the twenty-first century shift toward imagining cities in toto as vast urban laboratories, Hughes demonstrates how New York draws on its considerable cultural resources to construct an alluring new scientific imaginary by which to attract the critical resources of techno-cultural innovation. He examines the new Cornell Tech campus and the Hudson Yards redevelopment as demonstrations of New York's unique brand of boutique, bespoke digital urbanism. Hughes argues New York conjures a speculative scientific milieu in which posture and projection are privileged over empirical merit.

As the last three chapters in this volume make abundantly clear, the migration of the laboratory experiment outside of its formally contained spaces, into alternate ecological systems, cities, and interstellar utopias, foregrounds an even greater expansion and augmentation of its raison d'être. An exponentially warming planetary atmosphere renders the biosphere itself the experimental subject as scientists now contemplate its geoengineering and bioengineering. An experiment of unprecedented scale, complexity, and criticality, indeed it is the entirety of the world and the aggregate of its contentious and profligate lifestyles that now constitutes a global laboratory. This is perhaps the greatest experiment and social challenge of all time. In this last respect, we hope our book helps illuminates the "darker, hidden resources of the quotidian dimensions of organizational [scientific] life," a *science of imaginary solutions* to which all our respective fates are now inextricably wed.[6]

Notes

1. Bruno Latour and Peter Weibel, *Making Things Public: Atmospheres of Democracy*, catalog of the show at ZKM (Cambridge, MA: MIT Press, 2005).

2. Albena Yaneva and Bruno Latour, "Give Me a Gun and I Will Make All Buildings Move: An ANT's View of Architecture," in *Explorations in Architecture: Teaching, Design, Research*, ed. Reto Geiser (Basel: Birkhäuser, 2008), 80–89. She contributed chapters to two books coedited by Latour; see Albena Yaneva, "A Building Is a Multiverse," in *Making Things Public*, ed. Bruno Latour and Peter Weibel (Cambridge, MA: MIT Press, 2005), 530–535, and Albena Yaneva, "Challenging the Visitor to Get the Image," in *Iconoclash, Beyond the Image War in Science, Religion and Art*, ed. Bruno Latour and Peter Weibel (Cambridge, MA: MIT Press, 2002), 421–422.

3. Bruno Latour and Steve Woolgar, *Laboratory Life: The Construction of Scientific Facts* (Princeton, NJ: Princeton University Press, 1986), 32.

4. Jonas Salk, "Introduction" (1979), in Latour and Woolgar, *Laboratory Life*, 12.

5. Steven Shapin, *Never Pure: Historical Studies of Science as if It Was Produced by People with Bodies, Situated in Time, Space, Culture, and Society, and Struggling for Credibility and Authority* (Baltimore, MD: John Hopkins University Press, 2010).

6. Philip Hancock, "Uncovering the Semiotic in Organizational Aesthetics," *Organization* 12 (2005): 30.

1   Lifestyle Science: Its Origins, Precepts, and Consequences

Russell Hughes, Sandra Kaji-O'Grady, and Chris L. Smith

That the experiment is not a neutral space that excludes the presence, expectations, prejudices—indeed, the sensual apparatus of the scientist—is one of the strongest claims made by sociologists of science. It is the central argument of Bruno Latour and Steve Woolgar's *Laboratory Life: The Social Construction of Scientific Facts* (1979) and has become widely repeated in science and technology studies (STS). Yet, among scientists who maintain faith in the objectivity of the scientific method, the idea of the social construction of science, rooted in the socialization of scientists, is only provisionally accepted. In a parallel manner, there has always been a tension between the laboratory as a structure *outside* of the experiment, and the lab as a subject of experimentation. Hence, the paucity of literature on the architecture of laboratories, even in STS where the *placelessness* and neutrality of the laboratory have been strongly contested. The foundational mythology of the laboratory, Thomas Gieryn suggests, might make "science-buildings into a 'hardest case' for demonstrating that space and place are fundamentally involved in the reproduction of social life."[1] So while the geographical location of laboratories has been reinserted in historical considerations of science, we hear little about their spatial planning and circulation, furnishing and interior decor, internal and external views, constructional system, volumetric composition, facade treatment, continuity or discordance with existing urban or landscape settings, representational ambitions, and stylistic expression. Indeed, the extraordinary architecture of the Salk Institute for Biological Studies goes unmentioned in the English edition of Latour and Woolgar's *Laboratory Life*. Constructed in 1965, after a long period of gestation and emendation by its architect Louis I. Kahn, the Salk Institute is a site of architectural pilgrimage and tourism. The iconic plaza of the Salk Institute is described grandiosely by architectural historian Herbert Muschamp as "the most sublime landscape every created by an American architect"[2] and by critic Paul Goldberger as "the greatest outdoor room in American architecture since Thomas Jefferson's Lawn,"[3] but in *Laboratory Life* there is no mention of its courtyard. While Salk himself objected to the fetishization of the plaza, claiming "the building can't be read from the court," in Latour and Woolgar, the building can barely be read at all.[4] Discussions between scientists are noted as having taken place in an office, a laboratory, or a lobby. These flat denotations are linked to a crude plan diagram and are imbued with no architectural qualities whatsoever. Only in the later French translation of *Laboratory Life* (1986) did Latour, in a new preface, mention the building, misleadingly describing it as a mixture of "a Greek temple and a mausoleum."[5]

The Salk Institute's architecture was, and is, in fact, much more than a Greek temple, a mausoleum, a hilltop village, a medieval cloister, the Taj Mahal, a citadel, or any of the other structures to which it has been likened.[6] For Kahn, the architecture had to "convey a way of life."[7] The design encapsulates moral values of objectivity and individual discovery, an ethos of intellectual purity and the idea of science as a public good. What Kahn and Salk recognized was that efficiency and isolation had been surpassed as the key determinants of the laboratory's architectural expression. Architecture was to be engaged to recruit star scientists, inspire young people to embark upon careers in science, and attract philanthropy and industry partners. Architecture was to sponsor conversation and collaboration between scientists, enhance the performance and health of employees, and, thus, accelerate discovery. Architecture was charged with engaging the public in scientific endeavors, not only through spaces in which events for advancement and industry participation might be hosted, but also by communicating the importance (and content) of scientific research.

The idea of science as a public good, however, has taken quite a beating since the construction of the Salk Institute for Biological Studies. Confidence in scientific progress as a public project has faltered in the face of the evident self-interest and profiteering of the pharmaceutical industry, increasing dependency on private funding for research, the specter of genetic engineering and biological warfare, the politicizing of climate change science, and other evidence that science is, as Jean-François Lyotard reported in 1979, tightly interwoven with government and administration.[8] In *The Postmodern Condition*, Lyotard declares that "the games of scientific language become the games of the rich, in which whoever is wealthiest has the best chance of being right."[9] Scientific research is now resourced by and profits the powerful through novel relationships among government, universities, industry, and speculative investors. The assembly of transnational and private-public consortia and partnerships serve to gather enormous amounts of capital needed for research, and this ensures that discovery is linked to profit, markets, and the direction of resources by and for the powerful.

In this context, leading scientists and managers of scientific research enterprises have become savvy clients and consumers of architecture. Attention has shifted toward

Figure 1.1a, b

(a) The plaza of the Salk Institute for Biological Studies (1965). Photograph by Russell Hughes, 2017. (b) The sunken courtyards of the Salk Institute for Biological Studies with surfboard. Photograph by Sandra Kaji-O'Grady, 2015.

design as a tool for accelerating scientific discovery through the socialization of scientific communities. This has seen scientists and scientific managers, along with their architects, very much concerned with the configuration, expression, and aesthetic effect of laboratory design. Or, as Francis Duffy, whose practice DEGW specializes in the design of workplaces responsive to changes in management theory, writes, "design has become essential today for corporations using capital investment as a lever to affect organizational change."[10] Our concerns, though, extend even further than the idea of the laboratory as a site for managerial experiments in facilitating creativity and persuading the public of the value of experimentation. This book will come to suggest that the laboratory is a privileged site in the larger social experiment that has seen new knowledge economies, new vocational and leisure praxis, new modes of communication, and new technologies of self-care that are, by their very nature, co-extensive. While the scientist is constructed by the laboratory, they may be only the first step in a more generalized reconfiguring. Indeed, the laboratory presents us with multifarious new ways of being human, both in the mode of an experimental question *and* as a consequence of its experimental praxis.

Lifestyle Science

The salience of this book derives, in part, from the phenomenal investment in laboratory buildings in the last decade; among them are the Francis Crick Institute in London (2016) for USD 886 million (GBP 700 million), Weill Cornell Medical College's Belfer Research Building (2014) in New York for USD 650 million, and the Victorian Comprehensive Cancer Centre (2016) in Melbourne for USD 754 million (AUD 1 billion). These centers of biomedical research benefit from public and private interest in cancer, neuroscience, and anti-aging.[11] There are also nonmedical fields that attract generous funding and that have seen the construction of flagship laboratory buildings; for example, the elegant stone-clad Sainsbury Laboratory (2010) for research into the genetic modification of food plants in Cambridge, and the National Graphene Institute (NGI) (2015) in Manchester, England, for research into applications for graphene. Where the NGI was funded by the UK government and the European Union's European Regional Development Fund, the Sainsbury Laboratory was paid for through the charitable foundation of Lord Sainsbury of Turville, an heir to the eponymous supermarket chain and a former Labour minister. Designed by Stanton Williams, and constructed in 2010 at the relatively modest cost of USD 103 million (GBP 82 million), it won the Royal Institute of British Architects (RIBA) Stirling Prize for the best British building in 2012. Critic Oliver Wainwright claimed the architects had "recast what might have been an anonymous prefab shed, housing a functional stack of labs and pipes, into nothing short of a temple to botany."[12]

From the largest and costliest new laboratory buildings, to those with smaller budgets, can be seen attempts to create public spaces for persuasion and education, to represent scientific ideals and content through architecture, and to create lifestyle facilities to attract, retain, and motivate scientists. In the last decade, the lifestyle aspect of the architecture of scientific research has penetrated all levels and locales. The shift away from the self-effacing laboratory buildings of the mid-century to today's "temples" may, at first, seem to simply reflect changing stylistic commitments specific to the architectural discipline, which are now in reach of this building type as construction budgets have grown. Or, this shift may seem to be merely imitating the ostentatiously youthful workplaces of the tech industry, with their ping-pong tables and scooters. In fact, the situation is much more complex than either of these explanations. Richly appointed laboratory architectures, and the lifestyles they spatially and symbolically foster, signal a seismic shift in both the role of the scientist and, more specifically, what constitutes scientific practice. Science has adopted the operational principles of corporate organizational management into its own regimes of production to garner private support and funding. As a consequence, discourses of "vocational leisure" that saturate the corporate and entrepreneurial sector have migrated en masse into sites of scientific research. The creativity, and entrepreneurship such amenities are presumed to facilitate have been key aspects of intellectual work for decades, as we will discuss. Creativity is no longer an expression of individuality; rather, it is organizationally nurtured through managed social relationships.

In the case of laboratories, while their history dates back to the seventeenth century, the effect of social and spatial relations on scientific discovery emerges as a subject of concern in the 1930s. The historian of science, Peter Galison, traces the idea of scientific socialization to the period when European theoretical physicists fleeing Nazi Germany made their way to the United States and found themselves working with more pragmatic Americans.[13] According to Galison, "The war changed all the rules. At places like Los Alamos and Oak Ridge and MIT's RadLab [sic], you had mathematicians and theorists literally sitting on the other side of the desk from engineers. It was transformative."[14] During World War II, the need to accelerate technological advances prompted the allied forces to swiftly overcome the cumbersome divisions between science and engineering, and military and civilian research.[15] MIT, a major beneficiary of National Defense Research funding to pursue these ends, spawned the legendary Radiation Laboratory or "Rad Lab." Housing an unusually large number of research projects, from radar technologies to long-range navigation and the aiming of anti-aircraft guns, the Rad Lab's crammed, haphazard spatial arrangement incidentally facilitated fortuitous chance encounters among scientists from different backgrounds who could query each other's assumptions and, in turn, collaborate on what were then new "interdisciplinary" approaches to scientific problems.[16] In short, the discoveries made during this tumultuous period at the Rad Lab, Los Alamos, and Oak Ridge advanced techno-scientific knowledge considerably, and popularized the notion that "chance encounters," "happy accidents," "lateral thinking," and "interdisciplinary inquiry" were key to the acceleration of techno-scientific discovery and production.

The pseudo-science of "Space Syntax" has been invoked to give an empirical legitimacy to the rhetoric around the design of informal social spaces in research buildings. This nondiscursive theory of architecture has had a special resonance in science because of Bill Hillier and Alan Penn's attempts in the 1980s and early 1990s to quantify the impact of spatial organization on the behavior of laboratory occupants using its methods.[17] Hillier and Penn sought to numerically quantify the impact of co-location, density, and proximity between scientists on the frequency of their face-to-face interactions. Their argument, that discovery could be accelerated through design, overreached the data. In any case, the social experiment that is the contemporary laboratory is much wider in its scope, techniques, and effects than co-location or forcing everybody to use the same stairwell.[18] Space Syntax methodologies do not embrace the subjective, atmospheric, and affective powers of design, because these qualities cannot be easily measured. Yet, architects, developers, funding bodies, and managers of scientific organizations have embraced the intangible and sensual, and have done so with gusto. The leaders of scientific institutions recognize the full scope of the designer's vocabulary, from squashy chairs to landscape views. They link it to their own ability to manipulate other areas of sociality, from commensality to extracurricular bicycle rides for charitable fundraising.

To understand the proliferation of lifestyle science and its resort-style laboratories, we need to draw on a broader theoretical discussion of lifestyle and labor, life and science. This introductory chapter will outline three prominent historical and theoretical discourses that inform and account for the lifestyle laboratory phenomenon examined herein: first, the concept of lifestyle as a practice of conspicuous consumption and leisure; second, the shaping or production of an entrepreneurial self in the context of new forms of labor management that seek to harness creativity and knowledge; and third, the aestheticization of laboratory environments as a project of integrating and reproducing scientific knowledge in an economy of financial speculation. The theorists this chapter turns to—Thorstein Veblen, Pierre Bourdieu, Michel Foucault, Maurizio Lazzarato, and Félix Guattari—did not "follow scientists around," as Latour and Woolgar did.[19] They did not develop their arguments through the ethnomethodology of Latour and Woolgar, Karin Knorr-Cetina, Michael Lynch, and others who, over an extended period, closely observed the daily practices of scientists in the laboratory.[20] Their ideas need to be tested in the context of the institutions, ideologies, practices, and people that constitute the landscape of scientific research. While the chapters in this book do this in greater detail and approach the subject in diverse ways, here, theories of lifestyle as they pertain to science are introduced using three (unwitting) scientists. These three have been selected for their high personal and professional profiles, the generational shifts they embody, and the architectural specificity of the laboratories they inhabit, or haunt: James Watson (b. 1928), J. Craig Venter (b. 1946), and Nina Tandon (b. 1980).

Three Scientists

James Watson was appointed director of the Cold Spring Harbor Laboratory (CSHL), on the north shore of Long Island, New York, in 1968. It was here, fifteen years earlier, that he and Francis Crick made their first public presentation of the DNA double helix at a symposium entitled "Viruses." The pair received the 1962 Nobel Prize in Physiology and Medicine for the discovery of the structure of DNA. From its humble beginnings as a Fish Hatchery and Biological Laboratory in the 1890s, CSHL has become one of the world's most renowned not-for-profit private laboratories. It is focused on molecular biology and genetics with over six hundred researchers and technicians with expertise in cancer, neuroscience, quantitative biology, plant biology, bioinformatics, and genomics. The CSHL campus architecture is an odd conglomeration of styles that exercise a form of nostalgia. The architecture is an emulation, a collection of stylistic expressions that are geographically and historically misplaced. For five decades Watson has been instrumental in guiding the redevelopment and expansion of the CSHL's research activities and its grounds and buildings. He is now the CSHL's chancellor emeritus and remains a long-standing resident on the campus. Watson has written several accounts on the discovery of DNA and his life in science: *The Double Helix: A Personal Account of the Discovery of the Structure of DNA* (1968); *Genes, Girls and Gamow: After the Double Helix* (2001); *Avoid Boring People: Lessons from a Life in Science* (2007); and *Father to Son: Truth, Reason and Decency* (2014). An eccentric and often divisive figure, Watson has been called many things from the "Einstein of Biology" by Max Delbrück to the "Caligula of Biology" by Edward O. Wilson.[21] Indeed, Wilson once claimed that Watson was the most unpleasant human being he had ever met.

J. Craig Venter, on the other hand, a long-time rival of Watson's, has been equally divisive. He developed the "shotgun" sequencing method that led to the decoding of the first human genome, not incidentally extracted from his own sperm. Venter also "created" (computationally reconstructed) the world's first wholly synthetic organism, an act that saw him accused of "trying to short circuit millions of years of evolution and create his own version of the second Genesis."[22] He has been compared to God and Hitler in equal measure[23] and is considered "one of the few great rock stars in life sciences."[24] Where Watson's affiliation with CSHL has been steady, Venter's business activities are dynamic. The J. Craig Venter Science Foundation was launched in April 2002, merging three of the five not-for-profit research companies Venter had previously established. He personally gave the foundation a USD 100 million-plus endowment that he had amassed from a previous venture, Celera. In 2005, Venter launched a for-profit company called Synthetic Genomics, with the help of venture capital. He remains its chairman and co-chief scientific officer.[25] Venter is unusual in having been able to establish in his lifetime a purpose-built research laboratory bearing his name. The three-story headquarters of the J. Craig Venter Institute (JCVI) in La Jolla, California, designed by Zimmer Gunsul Frasca, opened in 2013 at a construction cost of USD 48 million. Like Watson, Venter has published accounts of his life and scientific discoveries, most recently *Life at the Speed of Light: From the Double Helix to the Dawn of Digital Life* (2014). His earlier biography, aptly described as a "study in ambition," was titled (without irony), *A Life Decoded: My Genome, My Life* (2007).[26] In his review of the book, Steven Shapin describes Venter as "aggressive, arrogant and ruthlessly competitive" as well as "belligerent, innovative, ambitious and entrepreneurial."[27]

Venter's marriage of business and science in the 1990s was novel, but for younger scientists such as Nina Tandon it is an established and necessary path. Tandon's PhD research focused on electrical signaling in the context of tissue engineering. While a postdoctoral researcher in the Laboratory for Stem Cells and Tissue Engineering at Columbia University, she met Sarindr Bhumiratana. The two founded EpiBone in 2013, a startup company researching the use of adult stem cells to grow bone cells outside of the body for personalized, living bone grafts. The company has been profiled in the *New York Times*, *Huffington Post*, and *Scientific American*, and the work has been the subject of TedMed, and segments on CNN and the BBC, among a host of other media outlets. Tandon was one of Fast Company's 100 Most Creative People in Business (2012), named CNN's Tech Superhero (2015), one of Crain's '40 Under 40' (2015), was chosen as one of the World Economic Forum's 2015 Technology Pioneers, and is a three-time TED speaker. EpiBone was a founding tenant of a co-share city-sponsored lab space in Brooklyn, furnished with all the hallmarks of gentrified hipster-chic. It even has its own "artist in residence." Tandon is too young to have written her biography, but her engagement with both formal and social media means much is publicly known about her interests, opinions, and activities beyond the laboratory. *Crain's New York Business* magazine summarizes her thus: "Nina Tandon eats lunch at the desk, has an iPhone, wears high heels and a miniskirt, is highly organized, [and] uses social media once a day."[28]

In the three sections to follow, these three scientists will each be deployed to illustrate and capture theoretical arguments around lifestyle science and the shifts that laboratories have at once marked and, indeed, instigated.

Lifestyle as a Practice of Conspicuous Consumption and Leisure

The relationship between work and leisure first emerged as a topic of sociology in Thorstein Veblen's 1899 analysis of affluent American societies, *The Theory of the Leisure Class*.[29] Veblen famously and controversially identified a "business" or "leisure class" that profited through economic predation upon the productive class of workers. Veblen proposed that leisure itself held social value because it portrayed the absence of a need to work and was, thus, a sign of capitalist ingenuity and entrepreneurial acumen. In particular, Veblen's analysis looked at how the wealthy managed to maintain and accrue money, not through "work," but through careful financial investments made as a result of contacts forged in elite social and leisure settings.

Throughout the course of the twentieth century, numerous sociologists, from Max Weber to Georg Simmel to Bourdieu, studied the use of leisure as a signifier of social position.[30] Forms of leisure activity have been convincingly shown to convey social class or status.[31] Thus, there is significance in the fact that Watson collects art, plays tennis, drives his Jaguar XJL around the North Shore's country roads, and enthusiastically attends fund-raising galas and dinners with New York's elite financiers and philanthropists. His wife, Elizabeth, a graduate of Radcliffe, a private women's liberal arts college now part of Harvard University, describes herself as a "happy hostess" and sits on numerous boards for museums, botanic gardens, and historic preservation.[32] Venter's leisure pursuits, on the other hand, are of a different shade, and the popular press has eagerly followed this "x-surfing" vagabond and his various exploits. As told by (or to) *The New York Times*, these include being out on the deck of his ninety-five-foot sailboat in a gale, "riding his German motorcycle through the California mountains, cutting the inside corners so close that his kneepads skim the pavement," and "snorkeling naked in the Sargasso Sea surrounded by Portuguese men-of-war."[33] The *Wall Street Journal* reports that in addition to owning a "gas-guzzling" Range Rover and an Aston Martin, and having a penchant for rare vintage motorcycles, Venter enjoys "doing large donuts" in his "45-foot jet boat."[34] Venter's third, and current, wife is his publicist Heather Kowalski, which may, in part, explain why he is accused of "science by press release."[35]

Meanwhile, on EpiBone's Internet homepage, Tandon describes herself as enjoying "yoga, rock climbing, surfing, and growing a strong business while growing strong bone."[36] She makes jewelry and is well networked in the art, design, and architecture scenes.[37] In August 2016, Tandon married technologist Noah Keating (his company Mathbeat Industries blends mobile device experiences, live events, and interactive installations). The ceremony took place at New York City's Louis Kahn-designed FDR Four Freedoms Park. The pair defied convention by donning top-to-toe black for the ceremony, while for the reception, Tandon, whose ethnicity is Indian, wore a sari, and Keating a kurta. Tandon's Pinterest board is almost entirely made up of photographs of interiors, furniture, design products, and architecture.[38] Where Watson is old-school East Coast establishment, and Venter a risk-taking Baby Boomer with perpetual stubble and a tan, Tandon belongs to a new flexible class of techno-scientific entrepreneur.

Tennis, sailing, and yoga are no doubt *experienced* as personal leisure choices for these three scientists, but, as Pierre Bourdieu would argue, they are over-determined by the milieux these individuals inhabit or to which they aspire. Bourdieu argues in *Distinction* (1984) that if the deployment of tastes in everyday life reproduces social class boundaries, then it is plausible to breach those boundaries through the appropriation of the material and cultural signifiers of taste.[39] Veblen, too, had observed that inherited wealth or an elite occupation does not in itself serve as admission to the upper classes—access depends on the adoption of an acceptable set of values and lifestyles. Conversely, one might successfully assert upper-class identity, as the Watsons convincingly do, without the requisite income or assets. Indeed, the differences in the Watsons' earnings and assets and those of their wealthy neighbors, some of whom also happen to be CSHL board members and benefactors, are discordant to the extreme. Francis Biondi, whose wife Jamie Nichols was elected chair of the CSHL board in 2010, is worth USD 1.64 billion.[40] Louis Moore Bacon, one of the CSHL's most generous donors, is reputedly worth USD 1.81 billion.[41] According to the CSHL's submission to the Internal Revenue Service in 2012, Watson's salary as its chancellor emeritus was USD 384,238.[42] Bacon and Biondi each earn more annually than the entire payroll of the CSHL's 1,256 employees.[43]

Watson is ambivalent about the trappings of a wealthy lifestyle, or at least its capacity to distract from science. For example, Watson contemptuously describes his former collaborator, Crick, as having "wasted his time" at the Salk Institute after the two parted ways. Crick, Watson claims, "had a great place in La Jolla, a white Mercedes and he knew a lot of girls, so he had a good life. He was much brighter than me, but not as sensible."[44] Watson argues that his own pursuit of a "good life" is in the interests of the organization. Interviewed by the *New York Times*, he claimed, as a manager of a scientific research institution, "You have to like people who have money. I really like rich people."[45] He elaborates why in *Avoid Boring People: Lessons from a Life in Science* (2007). He states:

Research institutions must have rich neighbors nearby who are inclined to take pride in local accomplishments. ... Entering worlds where your trustees relax—joining their clubs or vacationing where they go with their families in the summer, for instance—is a good way to put relations on a social footing. Seeing you as more friend than suppliant will incline them to go the extra distance for you in a pinch.[46]

From the earliest days of Watson's directorship, professional fund-raising help—as commonly used in other institutions—was eschewed in favor of a more personal approach to "potentially generous neighbors."[47] Long Island's wealthy residents, who make up the majority of the CSHL's board and donors, value membership on the boards of nonprofit organizations, volunteering, and visible gift giving to prestigious institutions. Philanthropy binds this social group, along with a range of interests and activities of the region—tennis, sailing, polo, prestige cars, collecting art, and houses.

What one does with leisure, and with whom, serves to stratify the social domain. What one purchases, and how this carries symbolic capital, interested Veblen also. Early in the twentieth century, corporations sought to distinguish themselves in an increasingly crowded marketplace through the stylization of ordinary everyday products. Stylistic features on new models and brands were loaded with cultural signifiers to make them more attractive and desirable. The most successful were those products imbued with "luxury branding" that appealed to consumers who, likewise, sought to distinguish themselves from others.[48] This desire enabled producers to demand higher prices than other products with comparable functions, leading to what Berkowitz et al. describe as "prestige-pricing,"[49] and the popularization of Veblen's earlier term "conspicuous consumption."[50] In this sense, selves are constructed by a set of associations and significations that are often summarized as "lifestyle."

According to linguist Geoff Nunberg, the term "lifestyle," though originating in the early twentieth century, gained popular use in 1972 with the rise in profile of Ralph Lauren's 'Polo' brand.[51] As Nunberg states: "Already interested in promoting a lifestyle with his ties, Ralph Lauren name[d] his line after a sport that embodies a world of discreet elegance and classic style: Polo." Lauren suggests of the concept, "We were the innovators of lifestyle advertisements that tell a story and the first to create stories that encourage customers to participate in that lifestyle."[52]

This "coating of physical products with symbolic content" underwent a subsequent shift in the 1980s as companies started marketing their products with traits such as "attitude" and "passion."[53] Style itself became a key criterion to fulfilling an individual's "life project," displayed or performed via an assembly of artifacts, practices, experiences, appearance, and bodily dispositions.[54] Products are purchased for their ability to express certain desirable lifestyle traits. In turn, celebrity purchases confer status on an object, and establish what has come to be known as "brand alliance." It is this understanding that sees Venter modeling for the prestige timepiece company Jaeger-LeCoultre, in a 2015 campaign that also features an athletic Argentinean polo player and the gamine filmmaker granddaughter of Charlie Chaplin.

Architecture, too, is a stylized product. The same ideas around associating oneself with objects saw Venter setting the architects of the new JCVI headquarters "the goal of designing the 'Salk Institute of the 21st Century.'"[55] Pursuit of the ripple effect of mutually reinforcing prestige led fashion company Prada to commission the Swiss architects Jacques Herzog and Pierre de Meuron to design Prada's Tokyo store in 2000. This was five years before the pharmaceutical company Actelion appointed Herzog and de Meuron to design their laboratory and headquarters in the architects' hometown of Basel. In Basel, too, rival pharmaceutical company Novartis has assembled a collection of buildings by Pritzker Prize Laureates Frank Gehry, Tadao Ando, David Chipperfield, and SANAA. The same desire for status prompted the real estate and development company of the Singapore Government, JTC, to commission Zaha Hadid to master plan its "one-north" science and technology campus. Yet, architectural prestige is conferred in many ways, not just through the reputation of the architect. Overseeing the development of the CSHL campus since the 1970s, Watson's commissioning of new buildings from Centerbrook architects, in a range of historically-inspired domestic styles, is a critical component of his social climbing. Centerbrook's most recent additions to the CSHL's Hillside Campus were described by a *New York Times* critic as an "architectural sleight of hand [disguising] the new labs as a miniature Bavarian hilltop village."[56] The approach Watson has pursued at the CSHL, eschewing fashion and signature architecture for a retrogressive vernacular, has made it difficult to accommodate changing technologies of construction, servicing, and programming, hence the subterranean and awkward laboratories of the Hillside Campus.

Watson is not put off by the juxtaposition between contemporary science and the architecture of a preindustrial era, nor is he shy about imposing his own taste preferences. He lives in a pink and cream reproduction Georgian house designed by Centerbrook and built in 1994.[57] Allegedly, it is "inspired by the classic farmhouses outside Venice designed in the late 16th century by the Italian architect and author Andrea

Figure 1.2
Frank Gehry's contribution to the architectural collection of Novartis on its Basel campus. Photograph by Sandra Kaji-O'Grady, 2014.

Palladio."[58] It sits at the most northern boundary of the campus, facing the outer harbor, with acres of grass and trees between it and the laboratories. The Watsons commissioned a painting by Bill Jonas, of the view of the harbor from the site occupied by the house before the laboratory was founded "because there were none in existence."[59] *Whaling Vessels in Pilgrim Lake* is in the realist style of eighteenth-century American naval art. By comparison, Venter and his wife live in a USD 6 million renovated cliff-top home of white walls and expansive glass overlooking the Pacific in La Jolla, "with sweeping curvilinear architecture" and a "wine room that doubles as a walk-in humidor."[60] The house features a large sculpture of a whale carved from a timber burl.[61] Tandon resides in a duplex on Roosevelt Island with a floor of recycled longleaf pine from the demolished Domino Sugar Factory, shelves filled with Le Creuset bakeware, heather-gray bed linen, and a portrait of Frank Ape doing yoga by local street artist Brandon Sines.[62]

Though preferences in architecture, antiques, art, food, sports, or cars may appear superficial and unrelated to scientific research, these selections reflect and communicate the alignment of the scientist and his or her laboratory with the social and financial milieu responsible for its funding and governance. They function as a further means toward assuring operational (and career) success. As a result of the near-wholesale shift from public to private funding in recent years, science has adopted the operational principles of corporate organizational management, in particular the "aestheticization" of vocational space, as a means to attract the venture capital so critical to its operation. This back-and-forth migration between science and organizational management culminates in the deployment of leisure and lifestyle tropes and conventions in scientific settings as a means toward further attracting venture capital(ists), increasing employee productivity, and, in particular, increasing the quantity *and* quality of vocational yield.

The Entrepreneurial Self

The project of the self, or what Paul du Gay identifies as the emergence of an "entrepreneurial self," is one in which individuals are engaged in a process of perpetual self-actualization that is motivated by the desire to forge a successful career.[63] This concept is further developed by Doris Eikhof, Chris Warhurst, and Axel Haunschild, who suggest "work and life are [becoming] intertwined, even amalgamated, so that they [professional individuals] cannot or do not want to distinguish and disentangle work and life."[64] Leisure pursuits in and around the workplace become symbols of personal growth. Astrophysicist and early Internet pioneer Larry Smarr, for example, who founded and directs the California Institute for Telecommunications and Information Technology (Calit2), encountered this when he moved to California in 2000. Arriving from the American Midwest, "the epicenter," as he likes to say, "of the [U.S.] obesity epidemic," the then-overweight Smarr became immersed in Southern Californian fitness culture.[65] He recalls feeling extreme pressure to get into shape. This experience is a testament to how scientific success is now contingent upon engaging a broader suite of lifestyle assets and practices that range from self-control (Smarr), to recklessness and self-confidence (Venter), or self-discipline and personal resilience (Tandon).

The idea that scientists, individually and as a community, are made through practices of life and work resonates with the critique of biopolitical governance developed by Michel Foucault.[66] Foucault argues that selves are not pre-given but socially constructed in and by communities, institutions, logics, and lifestyles. In Foucault's understanding, success is a measure of one's proximity to the power structures of the moment, one's resonance with the *episteme*. The same is true of scientific discoveries. Foucault's teacher Georges Canguilhem suggests that science is not a matter of fact or fiction, but rather a matter of being operative or not: "in the truth" [*dans le vrai*] or not.[67] Foucault developed his ideas about biopolitics with direct reference to the architecture of the clinic and to Jeremy Bentham's eighteenth-century design for a prison. Bentham called his invention a Panopticon after the Greek myth of the giant Panoptes who, with a hundred eyes, was a very effective watchman. Bentham described the Panopticon, with its watchman concealed from the view of the prisoner, as a "new mode of obtaining power of mind over mind."[68] Foucault called this generalizable form of disciplinary power "panopticism," recognizing its applicability in the service of a school, workshop, or penitentiary. Foucault's use of Bentham's 1791 architectural model preempts much of the contemporary discussion of self-governance.[69] While Smarr, for example, was not overtly under the surveillance of an institution, the shame and inadequacy he felt in the face of athletically inclined colleagues led to exactly the sort of regulation of the self that interested Foucault. Indeed, Smarr has embraced the commercial and research opportunities offered by internalized surveillance. He has engaged in a decades-long project of quantifying his own health through daily tracking of his blood and stool biomarkers, and is vociferous in his advocacy of "4P" (predictive, preventative, personalized, and participatory) medicine.[70] Considered key to the construction of an identity of personal achievement and/or success, labor force self-monitoring is today an essential precondition of capital accumulation.[71]

This form of biopolitical self-governance is perhaps most overtly expressed in the now-ubiquitous organizational focus on health. Maintaining a healthy workforce not only translates into better productivity margins for organizations, it also offers an(other) opportunity for competition, where social stratification and hierarchical position can be further enhanced by "winning." Be it in the office gym, cross-fit class, early morning "team" bike ride or lunchtime jog (your pace or mine?), these "leisurely" activities create the opportunity for workers to socialize in a more informal setting. They afford the ability to exercise, quite literally, traits valuable for organizational productivity such as initiative, dedication, and control.[72]

EpiBone's Internet homepage, as previously mentioned, details their scientists' bodily investments alongside their academic qualifications. Tandon's athletic pursuits are analogous to her commitment to the success of the company she co-founded. Sending a message to extraneous parties, be they prospective investors, co-founders, or other employees, Tandon works hard at taking leisure seriously. Imagining that there is some form of essential or authentic self behind such images is beside the point. The issue is not whether she actually practices yoga daily, or if Venter really does speed on his motorcycle; rather, it is the community that is constructed in respect to such images that matters. Such examples work to induce a feeling of obligation or expectation for other employees that behavior modification is necessary for personal success and that of the company.

In this new "live-work-play" vocational milieu, the private realm of the self undergoes a wholesale colonization, wherein, according to Margo Huxley, "assorted agencies, authorities, and groups seek to shape and incite the self-formation of the comportments, habits, capacities and desires of particular categories of individuals."[73] Contemporary organizational management constructs a field of (constrained) possibilities that, though shaping many of the core practices of daily life, does not wholly prescribe or determine all of them. Instead, sufficient encouragement is given for the entrepreneurial employee to invent new and ideally better ways to perform, and thereby contribute to the sharpness of an organization's competitive edge. Individuality expressed through lifestyle is celebrated, so long as that expression is one that can be recuperated into productive and lucrative outcomes.

Tandon, furthermore, offers a very distinct illustration of self-transformation in the pursuit of the resources to support her research endeavors. Tandon completed an MBA so that she could, in her own words, transform "from a biotechnologist to a biotech leader."[74] Her transformation has been effected through the active recruitment of publicity, a shrewd understanding of the need to tell a compelling story about science, and the clever deployment of her status as a photogenic young woman in a sector dominated by aging men. This self-portrayal is knowingly and professionally managed. For TEDMED, Tandon promotes life as "an entrepreneurial journey."[75] Questioned about her future in science, Tandon claimed she was unconcerned about unemployment or risk, but rather "worried about the job I'm going to create! If you think like an entrepreneur, you're never going to be out of work."[76] What is the laboratory for such a journey? EpiBone launched with USD 350,000 from Breakout Labs, a program set up by PayPal co-founder Peter Thiel's personal foundation to "jailbreak" academic research into the market economy. Though considerable, the grant is not enough to sustain the project. While still at Columbia University, the EpiBone team received USD 10 million in federal grant money for its preliminary research, yet even with these resources did not reach the animal trial stage. Provided all goes well funding these and subsequent human trials, EpiBone's technology will, under a best-case scenario, reach the market in 2022 or 2023.[77] Significant funds will be required to keep the work progressing, all the while knowing that with this kind of experimental research comes a high risk of failure. When it was incorporated in 2013, Tandon cautioned her investors not to be hasty or expect a return, "We're slow and steady, we're science nerds, and we are aiming to help humanity."[78]

James Watson hosted gala dinners to convince New York City's establishment to fund cancer research. Allegedly, he untied his shoelaces before meeting prospective donors to convey the impression of a man preoccupied with higher thoughts.[79] It is a very different image of scientific nerdiness that the team at EpiBone present, one concomitantly crafted as cosmopolitan, urbane, flexible, multicultural, and youthful. EpiBone was the founding tenant of Harlem Biospace, a micro shared laboratory funded by the City of New York as part of a wider program to secure a broad ecosystem of bioscientific research—a program Hughes discusses at length in chapter 10. The hipster-chic laboratory environment of Harlem Biospace—from its uptown location to its recycled furniture—anticipates the stylistic predispositions of its future occupants. Indeed, its street-cred décor mirrors that of Tandon's home, and of Tandon herself.

While appearing self-determined and innovatively "in control" of her personal and professional destiny, Tandon, like all knowledge workers, is shaped by government policy and university degree programs aligned with market expectations. Tandon's public persona is, likewise, mediated through existing stereotypes as is abundantly clear in the photographic portraits of her in the mainstream media. In a challenging affront to her feminist politics and scientific credentials, Tandon was photographed in *Wired* magazine in the pose of the Bleeding Madonna, complete with beatific expression and *cœur sur la main*. Venter, by comparison, was photographed in his office for the *Harvard Business Review*, in a tradition of male thinkers that extends back to paintings of "Saint Jerome in His Study," seated with his loyal dog among the paraphernalia of a contemplative life. He is regularly depicted at the helm of his sailing sloop, *Sorcerer II*. Mediated through popular social and cultural channels, the scientist's life is a complex

Figure 1.3
Portrait of Craig Venter for the *Harvard Business Review* series "Life's Work" (2014). Photograph by Michael Lewis.

construction that, in the context of "applied science" involves the "application" of the scientist's ingenuity *outside* the laboratory's walls as much as within them.

The Economies of the Laboratory

In his memoir, *A Life Decoded* (2007), Venter recalls the race to sequence the genome at Celera, the privately owned company where he was president during the late 1990s. He remembers that morale "was not only high, it was electric," recalling "people were happy, excited and energized in a way I had never experienced before."[80] Venter speaks fondly of the subcultures that emerged within each team of researchers:

The team headed by Enge Myers and Granger Sutton fostered a true geek culture, complete with high-octane espresso makers, foosball, and ping-pong tables. ... Each Monday a battle would commence, when geeks, clad in plastic Viking helmets and armed with Nerf guns that shot foam balls and even the occasional inflatable mace, waged war on the bioinformatics group, who used Nerf crossbows as their weapons of choice. ... The cafeteria became the central meeting point where almost everyone ate, bonded, and brainstormed daily.[81]

Figure 1.4
Portrait of Nina Tandon for *Wired* magazine's January 2013 issue. Photograph by Evan Kafka.

Replicating the scene and success at Celera is the goal of every manager of a scientific research organization. Collaboration, passion, and participation have been key aspects of intellectual work for decades, but have come to be valued in what Lazzarato identifies as the organization and constitution of *all* kinds of labor as "immaterial labor."[82] What is proposed by architects and managers as the humanizing of the laboratory through spaces for conviviality is, in fact, the outcome of an assemblage in which, as Guattari claims, "man is no longer paramount."[83] It is the cafeteria, the games, and the weapons that operate on the "almost everyone." The production of the scientist may be a necessary step toward the productions of science; but both are collateral to the economies of the laboratory. In this sense, the architecture of the contemporary laboratory may be a visible point of condensation in what Lazzarato, after Guattari, refers to as an "a-signifying semiotics": "A-signifying semiotics are the semiotics of mathematics, stock quotes, money, business and national accounting, computer languages, the functions and equations of science. ... They do not represent anything. ... They function by powering and amplifying the assemblages in which these semiotics are at work. ... In a-signifying semiotics, sign flows work directly on material flows."[84]

In light of Guattari and Lazzarato, we propose the laboratory may be understood as the material manifestation of mechanisms of "powering and amplifying" and vice versa. The laboratory might construct the scientist but it is, itself, configured by mechanisms: financial tools such as hedge funds, venture capital, and mortgage rates; developers' profit margins; government incentives and tax breaks; labor costs, contracts, and regulations; the numbers, qualifications, skills, and attributes of graduates; the structure and availability of research and development funding; and the volatile markets for scientific knowledge and its application in commodities. Architecture emplaces and gives expression to these otherwise silent mechanisms. A close reading of one particular contemporary laboratory, the J. Craig Venter Institute, demonstrates this point.

The aforementioned "staged" projections of Venter's life as an adrenaline- (and by proxy, fossil) fueled maverick, conflict with his ambition for the JCVI to be the world's first net-zero energy laboratory. These contradictions are reconciled, however, by Venter's conviction that the survival of the human species, indeed life on the planet itself, is "100% dependent" upon *his* science for its future. As he states in a 2012 interview: "We need food, water, energy, fuel, and medicine for the world's growing population ... DNA is the software of life. ... We can engineer it to do what we want it to ... the key to understanding the future of life on the planet is learning how to write that software."[85]

The architectural expression of Venter's new laboratory speaks volumes about the way science now employs literally every trick, both old and new, to construct its persuasive, if paradoxical, agenda. ZGF's design borrows the raw teak and exposed concrete of the Salk Institute that Venter hoped to emulate, but has none of the formality or order with which Kahn imbued the earlier laboratory.[86] Where Kahn's design for the Salk foregrounded the offices of the lead scientists, it did so in a way that suggested their democratic engagement as a collective, with each office of equal size and prominence. Venter's office at the JCVI, on the other hand, is singularly large and revealed only from the rear of the building. It sits at the prow of the administration and facilities wing, and is the only office with ocean views. Every other room in the JVCI, regardless of its functional designation is, in effect, an antechamber, there to dramatize the procession the visitor or employee makes from entry to their audience with its namesake. Once in the office the full gravitas of "the man" is on display, with vintage motorcycles on plinths and a wall hung with a litany of awards and certificates (but not, yet, the coveted Nobel Prize). The visitor is confronted by two life-size models of Venter's brain, a bronze award in the style of Rodin's *The Thinker*, a flotilla of Charles Darwin paraphernalia, photographs with U.S. presidents, and an array of books, medals, trinkets, and a telescope. No experimentation is practiced in this overtly masculinized room, but Venter's take-no-prisoners approach to the business of science is apparent in the heavy-handed curation of every architectural and interior gesture, his reputation for reckless motorcycle racing a synecdoche for his race to (synthetically) save the planet.

Conclusion

As the chapters in this volume elucidate, science has been co-opted and recuperated by the world of organizational management. Through selected examples, from ordinary to extreme versions of the lifestyle tendency, the chapters in this book make explicit the connection between science and its places of consumption and production. The book does not distill any single conclusion, but rather it opens up a discourse about the multiple manners by which the laboratory constructs the scientist and the multifarious impacts of lifestyle on the generation of science. Consisting of a series of extraordinary tales from across the globe, brought to life with interdisciplinary insight and illuminating illustration, this volume delivers the architecture of experimental science into full view.

Beyond the cold logic of economic rationality delivered, legitimized, and doubled through the new species of lifestyle laboratory architecture identified herein, we also see these laboratories communicate the *illogical terra incognito* of twenty-first century science. Here the evolution of "life itself" (always liminal) is inextricably contingent upon the attitudes instrumental to the interpretation and shaping of this life. As depicted in Stephan Helmreich's exploration of the limits of biology (as experienced in the fields of artificial life, deep ocean microbial life, and astrobiology), today the question most confronting to science is not what life "is," but rather what it "was" in the past tense. Laboratorial practice informs, transforms, and now deforms the stable, referential, "natural," constant, or "given" we *understood* life to be.[87] In this post-natural, post-anthropocentric, indeed, post-postmodern milieu, it is this shadow of life cast by the deranged mediums of scientific interpretation that are most fascinating and perplexing. As Alberto Corsín Jiménez and Rane Willerslev argue, the limits of biology are "also the place where the concept [of biology] outgrows its shadow, and becomes something else."[88] Tentatively, this volume speculates as to whether the new species of lifestyle laboratory architecture—the very places in which contestations over what life was, is, and will be, take place—are constitutive of this limit. For it is these architectural forms, as experimental as the science they incubate, that cast biology's longest, most mutant morphological shadow.

Notes

1. Thomas F. Gieryn, "What Buildings Do," *Theory and Society* 31, no. 1 (February 2002): 46.

2. Quoted in Michael Crosbie, "Add and Subtract," *Progressive Architecture* 74, no. 10 (October 1993): 48.

3. Paul Goldberger, "Imitation That Doesn't Flatter," *New York Times*, April 28, 1996, http://www.nytimes.com/1996/04/28/arts/architecture-view-imitation-that-doesn-t-flatter.html?pagewanted=1 (accessed September 14, 2013).

4. Jonas Salk quoted in Esther McCoy, "Dr. Salk Talks about His Institute," *The Architectural Forum* 127 (December 1967): 32.

5. Henning Schmidgen and Gloria Custance, *Bruno Latour in Pieces: An Intellectual Biography* (New York: Fordham University Press, 2014), 28.

6. Analogies such as Sam Lubell's description of the Salk Institute as the "Brutalist Taj Mahal," and Karen Stein's "modern-day citadel" are retrospective. Sam Lubell, *The Mid-Century Modern Architecture Travel Guide: West Coast USA* (New York: Phaidon, 2016); Karen Stein, "Louis I. Kahn's Salk Institute Remains a Modernist Beacon," *Architectural Digest*, January 31, 2012, https://www.architecturaldigest.com/story/louis-kahn-salk-institute-la-jolla-california-article (accessed September 19, 2013). The comparison between the Salk Institute and monastic models, however, was promoted by Salk and Kahn. Salk had visited Assisi in Italy in 1954 and "conveyed to Kahn the idea that this is what I would like—the cloistered garden." Jonas Salk quoted in McCoy, "Dr. Salk Talks About His Institute," 31–32.

7. Louis I. Kahn to Jonas Salk, 14 Aug 1961, SP, MSS 1, Box 369, Folder 2, cited in Stuart Leslie, "A Different Kind of Beauty: Scientific and Architectural Style in I. M. Pei's Mesa Laboratory and Louis Kahn's Salk Institute," *Historical Studies in the Natural Sciences* 38, no. 2 (2008): 208.

8. Jean-François Lyotard, *The Postmodern Condition: A Report on Knowledge*, trans. Geoff Bennington and Brian Massumi (Minneapolis: University of Minnesota Press, 1993), 45.

9. Ibid., 45.

10. Francis Duffy and Jack Tanis, "A Vision of the New Workplace," *Site Selection and Industrial Development* 162 (April 1993): 428.

11. There remain, of course, scientific fields languishing in outdated, unglamorous accommodation, overlooked because of a perceived lack of direct economic impact. Poverty-related and neglected diseases, for example, "lack the necessary financial incentives" for attracting research support and are forced to rely on relevant charities. See Helen Ann Halpin, Maria M. Morales-Suárez-Varela, and José M Martin-Moreno, "Chronic Disease Prevention and the New Public Health," *Public Health Reviews* 32 (2010): 120–154, https://doi.org/10.1007/BF03391595 (accessed October 18, 2013).

12. Oliver Wainwright, "Why the Sainsbury Laboratory Deserved to Win the Stirling Prize," *The Guardian*, October 15, 2012, http://www.theguardian.com/artanddesign/2012/oct/15/sainsbury-laboratory-deserved-stirling-prize (accessed October 18, 2013).

13. Peter Galison, *Image and Logic: A Material Culture of Microphysics* (Chicago: University of Chicago Press, 1997), 243–291.

14. Spencer Reiss, "Frank Gehry's Geek Palace," *Wired*, May 1, 2004, https://www.wired.com/2004/05/mit/ (accessed March 11, 2007).

15. Fred Turner, *From Counterculture to Cyberculture: Stewart Brand, the Whole Earth Network and the Rise of Digital Utopianism* (Chicago: University of Chicago Press, 2006), 17.

16. Galison, *Image and Logic*.

17. Bill Hillier and Allen Penn, "Visible Colleges: Structure and Randomness in the Place of Discovery," *Science in Context* 4 (1991): 23–49.

18. Thomas Allen, *Managing the Flow of Technology* (Cambridge, MA: MIT Press, 1977).

19. Bruno Latour, *Science in Action: How to Follow Engineers through Society* (Cambridge, MA: Harvard University Press, 1987).

20. See Karin Knorr-Cetina, *The Manufacture of Knowledge: An Essay on the Constructivist and Contextual Nature of Science* (Oxford: Pergamon, 1981); Michael Lynch, *Art and Artifact in Laboratory Science: A Study of Work and Shop Talk in a Research Laboratory* (Boston: Routledge and Kegan Paul, 1985); and Michael Lynch, *Scientific Practice and Ordinary Action: Ethnomethodology and Social Studies of Science* (Cambridge: Cambridge University Press, 1993).

21. E. O. Wilson, *Naturalist* (Washington, DC: Island Press), 42.

22. Jeremy Rifkin quoted in Meredith Wadman, "Biology's Bad Boy Is Back," *Fortune Magazine*, March 8, 2004, http://archive.fortune.com/magazines/fortune/fortune_archive/2004/03/08/363705/index.htm (accessed November 11, 2011).

23. John Sulston and Georgina Ferry, *The Common Thread: A Story of Science, Politics, Ethics and the Human Genome* (London: Corgi, 2003), 176.

24. Bradley J. Fikes, "Biotech Cluster Gets Serious Star Power," *San Diego Union-Tribune*, October 5, 2013, http://www.sandiegouniontribune.com/business/biotech/sdut-craig-venter-biotech -science-institute-jolla-2013oct05-htmlstory.html (accessed December 2, 2015).

25. Craig Venter drew a salary of USD 708,462 from the JCVI in 2013. Venter's wife, Heather Kowalski, was paid USD 120,000 for "Public Relations Services." Form 990, Return of Organization Exempt from Income Tax, Department of the Treasury, Internal Revenue Service, J. Craig Venter Institute, 2013, http://www.guidestar.org/FinDocuments/2013/521/842/2013-521842938 -0aefb777-9.pdf(accessed December 2, 2015).

26. Georgina Ferry, "Learning the Lessons of Life," *Guardian*, October 27, 2007, https://www .theguardian.com/books/2007/oct/27/featuresreviews.guardianreview6 (accessed August 10, 2015).

27. Steven Shapin, "I'm a Surfer," *London Review of Books* 30 (2008): 5–8.

28. Irina Ivanova, "Nina Tandon, 35, EpiBone Chief executive," *Crain's New York Business*, http:// www.crainsnewyork.com/40under40/2015/Tandon (accessed December 2, 2015).

29. Thorstein Veblen, *The Theory of the Leisure Class: An Economic Study in the Evolution of Institutions* (New York: Macmillan, 1899).

30. Max Weber, *The Protestant Ethic and the Spirit of Capitalism*, trans. Talcott Parsons (New York: Charles Scribner's Sons, [1905] 1963); George Simmel, *The Sociology of George Simmel*, trans. Kurt Wolff (New York, The Free Press, 1950); Pierre Bourdieu, *Distinction: A Social Critique of the Judgment of Taste*, trans. Richard Nice (Cambridge, MA, Harvard University Press, [1979] 1984).

31. See Mike Featherstone, "Leisure, Symbolic Power and the Life Course," in *Sport, Leisure and Social Relations* (Sociological Review Monograph 33), ed. John Horne, David Jary, and Alan Tomlinson (London: Routledge and Kegan Paul, 1987), 113–138.

32. Elizabeth Watson, *Grounds for Knowledge: A Guide to Cold Spring Harbor Laboratory's Landscapes and Buildings* (Cold Spring Harbor, NY: Cold Spring Harbor Laboratory Press, 2008), 208.

33. Will S. Hylton, "Craig Venter's Bugs Might Save the World," *New York Times Magazine*, May 30, 2012, http://www.nytimes.com/2012/06/03/magazine/craig-venters-bugs-might-save-the -world.html (accessed December 4, 2015).

34. Sara Lin, "Craig Venter's Hangout," *Wall Street Journal*, March 12, 2010, https://www.wsj .com/articles/SB10001424052748704548604575098131101212908 (accessed July 14, 2014).

35. Sumit Singhal, "J. Craig Venter Institute in La Jolla, California by ZGF Architects," *AECCafe*, April 21, 2013, http://www10.aeccafe.com/blogs/arch-showcase/2013/04/21/j-craig-venter-institute -in-la-jolla-california-by-zgf-architects/ (accessed April 19, 2014).

36. "Our Team," EpiBone homepage, http://www.epibone.com/team/ (accessed May 7, 2017).

37. "Nina Tandon, PhD '09 BME," Graduate Student Affairs, Columbia Engineering, http://gradengineering.columbia.edu/nina-tandon-phd-09-bme (accessed April 14, 2017).

38. *Nina Tandon*, Pinterest, https://www.pinterest.com.au/ioanaru/tandon/ (accessed April 1, 2017).

39. Pierre Bourdieu, *Distinction: A Social Critique of the Judgment of Taste*, trans. Richard Nice (Cambridge, MA: Harvard University Press, 1984).

40. "Francis Biondi Profile," *Forbes*, https://www.forbes.com/profile/o-francis-biondi/ (accessed April 14, 2015).

41. "Louis Bacon," *Inside Philanthropy*, https://www.insidephilanthropy.com/wall-street-donors/louis-bacon.html (accessed April 9, 2015).

42. Schedule O, Form 990-PS. Return of Organization Exempt from Income Tax, Department of the Treasury, Internal Revenue Service, Cold Spring Harbor Laboratory, 2012.

43. In 2012 the CSHL employed 1,256 people and had salary costs of around USD 70 million a year; see *2013 Annual Report*, Cold Spring Harbor Laboratory, http://www.guidestar.org/FinDocuments/2010/112/013/2010-112013303-07d10dbd-9.pdf (accessed April 9, 2015).

44. Amanda Gefter, "Wilson vs. Watson: The Blessing of Great Enemies," *New Scientist*, September 10, 2009, https://www.newscientist.com/article/dn17771-wilson-vs-watson-the-blessing-of-great-enemies/ (accessed April 19, 2015).

45. Carol Strickland, "Watson Relinquishes Major Role at Lab," *New York Times*, March 21, 1993, http://www.nytimes.com/1993/03/21/nyregion/watson-relinquishes-major-role-at-lab.html?pagewanted=all (accessed April 14, 2015).

46. James Watson, *Avoid Boring People: Lessons from a Life in Science* (Oxford: Oxford University Press, 2007), 313.

47. Ibid., 286.

48. Peter Dobers and Lars Strannegård, "Design, Lifestyles and Sustainability: Aesthetic Consumption in a World of Abundance," *Business Strategy and the Environment* 14 (2005): 325–326.

49. Eric Berkowitz et al., *Marketing* (Homewood, IL: Irwin, 1992).

50. Veblen, *The Theory of the Leisure Class*.

51. Geoff Nunberg, "The Evolution of Lifestyle," *National Public Radio*, July 31, 2006, https://www.npr.org/templates/story/story.php?storyId=5594617 (accessed October 11, 2016).

52. "About Us," Ralph Lauren, http://global.ralphlauren.com/en-us/about/Pages/default.aspx (accessed March 29, 2014).

53. Elizabeth Hirschman, Linda Scott, and William Wells, "A Model of Product Discourse: Linking Consumer Practice to Cultural Texts," *Journal of Advertising* 27 (1998): 33–50.

54. Mike Featherstone, *Consumer Culture and Postmodernism* (London: Sage, 1991).

55. Ted Hyman quoted in ZGF Architects LLP, *J. Craig Venter Institute* online viewbook, http://issuu.com/zgfarchitectsllp/docs/j._craig_venter_institute?e=5145747/7966125 (accessed August 11, 2015).

56. J. Alex Tarquinio, "Long Island Laboratory's Expansion Hides in (and Under) Six Buildings," *New York Times*, June 23, 2009, http://www.nytimes.com/2009/06/24/realestate/commercial/24lab.html (accessed July 1, 2015).

57. Watson, *Grounds for Knowledge*, 127.

58. Ibid.

59. *CSHL News*, "Local Painter Brings 1848 Whaling Vessels in Cold Spring Harbor to Life," February 15, 2012, http://www.americantowns.com/ny/coldspringharbor/news/local-painter-brings-1848-whaling-vessels-in-cold-spring-harbor-to-life-8490802 (accessed August 11, 2015).

60. Hylton, "Craig Venter's Bugs Might Save the World."

61. Ibid.

62. Vivian Yee, "Salvaging a Long-Lasting Wood, and New York City's Past," *New York Times*, July 21, 2015, https://www.nytimes.com/2015/07/22/nyregion/salvaging-a-long-lasting-wood-and-new-york-citys-past.html (accessed July 1, 2015). The Le Creuset pan and Heather Grey Sheets are among items listed and purchased from Tandon and Noah's Zola wedding registry in 2016 (see https://www.zola.com/registry/ninaandnoah). Tandon's partner Noah Keating posted on Twitter his purchase of a small painting by Brandon Sines in November 2014, https://twitter.com/noha/status/529100523887415296/photo/1 (accessed July 1, 2015).

63. Paul du Gay, *Consumption and Identity at Work* (London: Sage, 1996).

64. Doris Eikhof, Chris Warhurst, and Axel Haunschild, "Introduction: What Work? What Life? What Balance? Critical Reflections on the Work-Life Balance Debate," *Employee Relations* 29 (2007): 325–333.

65. Mark Bowden, "The Measured Man," *The Atlantic*, July/August 2012, https://www.theatlantic.com/magazine/archive/2012/07/the-measured-man/309018/ (accessed May 12, 2017).

66. Michel Foucault, *The History of Sexuality: The Will to Knowledge*, vol. 1, trans. Robert Hurley (London: Penguin, 2008).

67. Georges Canguilhem, *Idéologie et Rationalité dans l'Histoire des Sciences de la Vie* (Paris: Librarie Philosophique J. Vrin, 1977); *Ideology and Rationality in the History of the Life Sciences*, trans. Arthur Goldhammer (Cambridge, MA: MIT Press, 1988), 13. Refer also to Michel Foucault, *Archaeology of Knowledge*, trans. Alan Mark Sheridan-Smith (New York: Pantheon Books, [1969] 1972), 224.

68. Jeremy Bentham, *The Works of Jeremy Bentham*, vol. 4 (Panopticon, Constitution, Colonies, Codification), Online Library of Liberty, Liberty Fund (1843), 39.

69. Michel Foucault, *Discipline & Punish: The Birth of the Prison* (London: Penguin, 1977), 195–228; see also Jeremy Bentham, *The Panopticon Writings*, ed. Miran Bozovic (London: Verso, 1995).

70. Larry Smarr, "Quantifying Your Body: A How-to Guide from a Systems Biology Perspective," *Biotechnology Journal* 7, no. 8 (2012): 980–991.

71. Lisa Adkins and Celia Lury, "The Labour of Identity: Performing Identities, Performing Economies," *Economy and Society* 28 (1999): 598–614.

72. Amanda Waring, "Health Club Use and 'Lifestyle': Exploring the Boundaries between Work and Leisure," *Leisure Studies* 27 (2008): 295–309.

73. Margo Huxley, "Geographies of Governmentality," in *Space Knowledge and Power: Foucault and Geography*, ed. Jeremy Crampton and Stuart Elden (Hampshire: Ashgate, 2007), 188.

74. Liz Welch, "How a Bone-Growing Startup Lured 66 Investors, Including Peter Thiel," *Inc. Magazine*, October 2015, https://www.inc.com/magazine/201510/liz-welch/blooming-bones.html (accessed November 11, 2015).

75. Zaina Awad, "Making a Living with Biology—Q&A with Nina Tandon," TEDMED, May 7, 2015, https://blog.tedmed.com/field-living-possibilities-qa-nina-tandon/ (accessed April 14, 2015).

76. Ibid.

77. "How to Regrow Your Own Bones," *Scientific American*, June 27, 2016, https://www.scientificamerican.com/article/how-to-regrow-your-own-bones/ (accessed May 12, 2017).

78. Welch, "How a Bone-Growing Startup Lured 66 Investors."

79. James Watson quoted in Mark Kuchner, *Marketing for Scientists: How to Shine in Tough Times* (Washington, DC: Island Books, 2012), 23.

80. Craig Venter, *A Life Decoded: My Genome, My Life* (New York: Viking, 2007), 261.

81. Ibid.

82. Maurizio Lazzarato, "Immaterial Labor," trans. Paul Colilli and Ed Emery, in *Radical Thought in Italy: A Potential Politics*, ed. Paolo Virno and Michael Hardt (Minneapolis: University of Minnesota Press, 1996), 133–147.

83. Félix Guattari, "Les Annees d'Hiver" (Paris: Les Prairies Ordinaires, 2009), 128, quoted in Maurizio Lazzarato, "'Exiting Language,' Semiotic Systems and the Production of Subjectivity in Félix Guattari," trans. Eric Anglès, in *Cognitive Architecture: From Biopolitics to Noopolitics*, ed. Deborah Hauptmann and Warren Neidich (Rotterdam: 010 Publishers, 2010), 512.

84. Ibid.

85. Craig Venter quoted in David Frost, "Craig Venter: 'The Software of Life,'" interview, *Aljazeera*, December 22, 2012, http://www.aljazeera.com/programmes/frostinterview/2012/12/20121219111158970847.html (accessed April 28, 2017).

86. Ted Hyman quoted in ZGF Architects LLP, *J. Craig Venter Institute, La Jolla Building Overview,* 2013,   http://www.jcvi.org/cms/fileadmin/site/sustainable-lab/JCVI-La-Jolla-Building-Overview .pdf (accessed May 14, 2015).

87. Stephan Helmreich, "What Was Life? Answers from Three Limit Biologies," *Critical Inquiry* 37 (2011): 671–696.

88. Alberto Corsín Jiménez and Rane Willerslev, "'An Anthropological Concept of the Concept': Reversibility among the Siberian Yukaghirs," *Journal of the Royal Anthropological Institute* 13 (2007): 527–544.

2 Beanbags and Microscopes at Xerox PARC

Kathleen Brandt and Brian Lonsway

Among the many mechanical and electronic instruments at the Computer History Museum in Mountain View, California, rests a vintage beanbag. It is situated quite ceremoniously on a plinth in front of a large photograph of six people sitting in similar beanbags in a nondescript room.

As a Computer History Museum exhibit, the beanbag serves as a symbol and synecdoche of the creative atmosphere in the early days of the Palo Alto Research Centre (PARC), the center famous for inventing such staples of modern computing as the graphical user interface, Ethernet, and the laser printer. Intentionally set on the opposite side of the continent from Xerox's corporate headquarters in Rochester, New York, PARC embraced the creative agency of a new breed of researcher/developer identified with the technology-embracing counterculture of the burgeoning Silicon Valley. The designers of PARC sought to reject the stodginess of a typical corporate research office in favor of a casual environment that embraced the "new" kind of creative thinking and collaboration required to fulfill the center's highly speculative mandate. Elsewhere at Xerox—even just down the hall from PARC's Computer Science Lab itself—researchers donned their lab coats, hunched over their data entry terminals, and leaned into their microscopes.

The (literal) institutionalization of the beanbag in the Computer History Museum recognizes both a particular moment in time and an ethos that has become increasingly pervasive since the early days of PARC. Through the age of startups and the dot-com generation, the geeks-in-a-garage "no-collar" culture of young innovators and "creatives" has become a cultural meme as much as a reality. The beanbag has reached the contradictory status of both an inside joke and a still-earnest symbol of creative corporate culture. We see proselytizers like Richard Florida, author of the widely influential *Rise of the Creative Class*, formalizing the ethos of *hipsterism*, even to the extent of hedging the betterment of our national productivity on the economic and political support of a culture of creativity.[1]

Figure 2.1
The PARC beanbag exhibit at the Computer History Museum in Mountain View, California.
Photograph by the authors.

The Computer History Museum exhibits the beanbag with a nod to both historical accuracy and cultural cheekiness, acknowledging that the transformative impacts of this commonplace piece of furniture in the domain of computing history are manifold. The beanbag functions ergonomically, culturally, and symbolically, and these various functions have been built upon, expanded, and in many cases reified since this famous placement in the early PARC. Our friend the beanbag has grown up, but not necessarily matured, in the hands of designers since its invention. Together with many allies including the fern, the ping-pong table, and the playground slide, the beanbag has become a de facto indicator and erstwhile instigator of creative capacity.

What is behind this evolution? And to what do we owe the iconic status of the beanbag? We need look no further than Bruno Latour and Steve Woolgar's observational work with the Salk Institute's scientists for a method to unpack the evolution of PARC's beanbags into the Googleplex. We see that designers, much like scientists, often like to create products that stabilize the ambiguities and contingencies of complex phenomena into what they, too, consider "facts"; statements that no longer "refer to the conditions of their construction."[2] What was once a "what if" statement about the *potential* of a particular arrangement of furnishings has become an accepted *reality* through what Latour and Woolgar call "microprocesses" in an "agonistic field," the production of authoritative documents, public and private debate, and discourse among professionals.

The variety of laboratory we are looking at here is a somewhat unique formation that focuses specifically on the enabling of a form of thinking directed toward the creation of "innovative" technological products. It is every bit a scientific laboratory as Salk, but privileges the "agonistic" work described by Latour and Woolgar that produces the ultimate "fact-products" (academic papers, in the case of Salk's scientists; designs both described and realized in the case of our laboratory's scientists) over the "benchwork." As such, we take the microprocesses inherent to these latter stages—and the forms of thinking that accompany them—as a central theme in our argument. We are looking, ultimately, at a quite different form of "fact-product"—in our case, laboratory designs themselves—but the arguments we appropriate from *Laboratory Life* remain the same.

Thinking

Thinking is among the more elusive processes that Latour and Woolgar identify as a key component in this "microprocessing of facts"—the numerous social and discursive events that scientists use to evolve and refine conclusory arguments from experiments. The laboratory is a place where, with a certain degree of autonomy from everyday life, scientific thinking uncovers (or constructs as the authors inform us) the known facts of our world. It is (literally) a place for labor, where thinking is serious work. However, thinking is hard for the ethnographer to document or study, leaving its contents and methods opaque. "It could be argued, for example, that the solitude of the individual scientist in thought excludes the sociologist by definition. Social factors are self-evidently absent from the activity of thought."[3] What goes on when we are thinking is notably difficult to articulate. Observers are typically only privy to the result, the "idea," that "aha!" moment, when thinking "results" in something that can be articulated. According to Latour and Woolgar, "Slovik proposed an assay but his assay did not work everywhere; people could not repeat it; some could, some could not. Then one day Slovik *got the idea* that it could be related to the selenium content in the water: they checked to see where the assay worked; and indeed, Slovik's idea was right, it worked wherever the selenium content of water was high."[4]

But Latour and Woolgar argue that it is critical to acknowledge the social dimensions of thought. In contrast to Slovik's idea, they note, "Watson's portrayal … does not situate himself in a realm of thought, but inside a real Cambridge office manipulating physically real cardboard models of the bases. He does not report having had ideas, but instead emphasizes that he shared an office with Jerry Donohue. When Donohue objected to Watson's choice of the enol form for picturing the bases, Watson pointed to actual textbooks of chemistry."[5]

While Watson situates his idea narrative in the material context of his office, Slovik simply arrives at his idea; one moment it was not there, and the next it was. By masking the contingent social processes within the thinking process that contribute to ideas, the "aha!" narrative privileges the same assumptions that divorce scientific facts from their social construction. Latour and Woolgar offer that "statements lie along a continuum according to the extent to which they refer to their conditions of construction," and in their argument "wish to show that the process of construction involves the use of certain devices whereby all traces of production are made extremely difficult to detect."[6] Invisible thinking processes neatly fit this claim: "the notion of someone having had an idea provides a highly condensed summary of a complex series of processes."[7] For the authors of *Laboratory Life*, these claims contribute to their groundbreaking arguments about the agonistic "constructedness" of scientific facts—of interest, one might hope, not only to future ethnographers, but future lab scientists as well. But what value might these arguments have not just for the people *in* the lab, but for the people who design them? While one might hope otherwise, the pretense of the "idea" formed without contingencies is, unfortunately, just as common in the design studio as it is in the laboratory.

Thinking, of course, is not just one thing: even in the Salk Institute lab, "depending on the argument, the laboratory, the time of year, and the currency of controversy, investigators will variously take the stand of realist, relativist, idealist, transcendental relativist, sceptic, and so on."[8] A rich body of literature specifically dedicated to understanding the thinking process, often specifically concerned with questions of creativity, has developed since the 1950s across many disciplinary channels; Herbert Simon's work on decision making, Edward de Bono's on thinking, and Seymour Papert's on learning, are but a few notable examples in the areas of education and organizational development.[9] Important contributors to the "data" that would underlie future work on the subject include the meticulous testing of creative thought by Don MacKinnon's Institute for Personality Assessment and Research undertaken since 1959,[10] as well as observational research on the impacts of LSD on creativity carried out by Myron Stolaroff's

International Foundation for Advanced Study between 1961 and 1965.[11] (Both of these were based in the Silicon Valley Area, conveniently close to PARC.)

So, if such thinking, with its vast range of modalities and high dependency on social and environmental contingencies, is to occur in the laboratory, how does a designer design for it? How can the complexity of the thinking process be made visible to designers such that they can support its range of modalities and meaningfully engage the social and environmental contingencies that lend it form?

The Social Construction of a Lab

This is a question not too far off from those asked in 1959 at the opening of Stanford University's Joint Program in Design (now the Design Impact Engineering program). A collaboration between the departments of art and mechanical engineering, the Joint Program was formed to embrace the potentials of "human-centered design," an approach that comprises interdisciplinary design methods that place human needs and desires at the center of design processes. Responding to the more mechanistic and science-like postwar approaches to design methodology that privileged efficiency and optimization, the human-centered approach began to develop research methods that would directly involve a project's key stakeholders in the decision-making process of designers.[12] It grew from the work of former MIT professor John Arnold, founder of the institution's "Creative Engineering Lab," who relocated to Stanford in 1957 with a joint appointment in engineering and management. Arnold was heavily influenced by many of the same systems theory, cybernetics, and information-technology ideas that inspired the aforementioned inquiries into thinking and creativity. He famously brought leaders in these very fields together for a summer MIT seminar.[13] At the time of the Joint Program's founding, this was a radical departure from traditional models of architectural and design teaching because it blended engineering design principles, artistic exploration, and managerial productivity. Together, these skills would prove invaluable for the substantial postwar transformations of the commercial organization as they sought to embrace new strategic decision-making processes and managerial styles, provoked by their acquisition of mainframe computers and their restructuring around new communication technologies. The incorporation of these technologies and the systems theory, cybernetics, and information theory-influenced processes into an organization's structure and identity required new ways of thinking about business organizations, and led to the substantial involvement of designers in the "new workplace."

In the realm of the physical work environment, from the office to the laboratory, one quite influential development that built on these methods was the *burolandschaft*, or "office landscape." A well-publicized 1967 Delaware office redesign for DuPont (then E. I. du Pont de Nemours and Company) by German workplace design consultancy Quickborner Team is commonly cited as its American introduction.[14] The *burolandschaft* sought to empower the new "knowledge worker,"[15] and challenge the rigid corporate hierarchies of the typical postwar office environment:

Office landscape planning began with a complete survey of all office staff and their individual patterns of communication. The layout was then developed around this matrix placing workers in close proximity to the other staff members they communicate with the most frequently regardless of status or departmental affiliation. In this way, office landscaping tossed out the American convention of relying on the prescriptive lines of communication as depicted by the organizational chart, and instead sought to uncover the real lines of communication.[16]

In large part because an organization's communication patterns (much like thinking processes) were not articulated and thus invisible to the casual observer, office landscapes often appeared as a "jumbled mess of desks arranged willy-nilly around the open space."[17] More precisely, they tended to be "characterized by a wide open space with clusters of desks arranged at different angles and the prominent use of curving screens and plants to demarcate the interior space."[18] The *burolandschaft* represented a new form of thinking about organization; the user-centered, rather than managerially determined, design process required its creators to conduct what design researchers now call "stakeholder research" and "experience mapping." Quickborner's field research collected and structured both personal narratives and official documents about workflows and communication process. It visually structured this "data" into relational diagrams and flowcharts. And it ultimately produced alternative office arrangements that explicitly laid visible the results of this research and design process.

Participant-engaged design and office planning; the alternative management arrangements; an increasing awareness of the value to managers of new modes of communication, collaboration, and creative thinking—these ideas were already becoming the convention rather than the exception. This was in large part due to a remarkable confluence of research on thinking, design, and creativity in the San Francisco Bay Area, as Rochester, New York-based Xerox selected Palo Alto, California, without fanfare, to house its new research center.

But the lack of fanfare was about to change, for as noted in the December 7, 1972 issue of the still-in-its-infancy *Rolling Stone* magazine, "here was [Xerox's] new multimillion-dollar research center spread out for unsupervised public view in a ratty rock music magazine, with actual Xerox scientists photographed in their t-shirts and jeans,

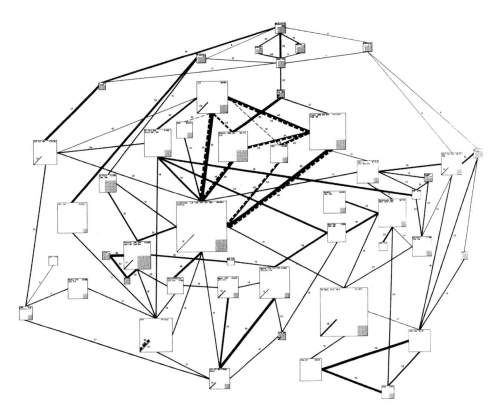

Figure 2.2
A map of communication connections and frequencies derived from user survey data used in the
burolandschaft planning process. Reproduced from John Pile, "Clearing the Mystery of the 'Office
Landscape' or 'Burolandschaft,'" *Interiors' Second Book of Offices* (New York: Whitney Library of
Design, 1969).

barefooted, lounging self-indulgently in beanbag chairs."[19] A now-legendary exposé
by *Whole-Earth Catalog* founder Stewart Brand and photographer Annie Liebowitz,
"SPACEWAR: Fanatic Life and Symbolic Death among the Computer Bums" was a tour
of Silicon Valley's computing research labs. Brand depicted them as full of long-haired,
bicycle-riding, video-gaming, counter-cultural hackers and hippies. The essay stands as
one of the more prominent and informative documents—an artifact—of the vanguard
computing culture of the Valley at the time. Xerox PARC was featured prominently,
much to the dismay of the suit-and-tie culture of its parent company. In the article,
Alan Kay, one of PARC's leaders and a visionary whose ideas informed many of today's
computing devices, described its employee profile, the "Standard Computer Bum," to
Brand in this way:

About as straight as you'd expect hotrodders to look. It's that kind of fanaticism. A true hacker is not a group person. He's a person who loves to stay up all night, he and the machine in a love-hate relationship. ... They're kids who tended to be brilliant but not very interested in conventional goals. And computing is just a fabulous place for that, because it's a place where you don't have to be a Ph.D. or anything else. It's a place where you can still be an artisan. People are willing to pay you if you're any good at all, and you have plenty of time for screwing around.[20]

Warmer climes and the proximity to Xerox's recent acquisition of a Los Angeles-based computer company were the rationales George Pake, PARC's first director, provided to Xerox headquarters. Pake also knew, however, that Palo Alto, home of Stanford University and center of the burgeoning Silicon Valley, was to be a profoundly important player in the next-generation digital technology development that was at the center of his mission. Plus, it was 2,700 miles away from Xerox headquarters, which gave PARC researchers a much-needed autonomy from the staid corporate power structures back east.

PARC opened in 1970, just a few short years before Latour's field work began at the Salk Institute to the south, and was on the cusp of an entirely new generation of lab typology. While AT&T's Bell Labs may be PARC's most obvious progenitor in the lineage of research laboratories—and likely much more of what Xerox was anticipating— it was of a markedly different era, dominated by an early-twentieth-century business culture of men in lab coats with sleeves rolled up and ties tucked into their blue or white collared shirts.[21] The Silicon Valley before PARC built its reputation quite literally on silicon, had its foundations in the form of the Shockley Semiconductor Laboratory, Fairchild Semiconductor, and Intel. These companies—among the country's earliest startups—were formed by ex-Bell Labs scientists, and inherited the culture of Bell Labs, even if with a more relaxed climate. The upcoming companies of Silicon Valley "were more informal and democratic in their organization and employee culture. ... They appeared to promote a human workplace not as a grudging concession to demoralized employees but as a valued asset to production."[22]

As applications of Silicon Valley's hardware components surfaced in the form of the microcomputer, the design and engineering of systems, assemblies, and retail products began to take center stage. It is here where the political significance of personal and connective technology drew the kind of figure represented by polymath Alan Kay, inventor of the programming language Smalltalk who joined PARC in 1970, along with their expansive, psychedelic, and mindful approaches to thinking. The home-computer revolution expanded from the Stanford computer laboratories into PARC and the fabled Palo Alto garage. It included both the hardware and software systems to make these devices not only comprehensible to, but also empowering of, the layperson.

The fact that the labs described in *SPACEWAR* were at the forefront of an entirely new technological project gave them a license, and specifically in the case of PARC, a mandate, to reinvent their work environment to accommodate their alternative modes of thinking. PARC was charged by Xerox's CEO to "prove the viability of the automated office by building functioning machines in sufficient numbers to be used, tested, and studied."[23] The PARC labs focused on digital and information technologies and did not operate like conventional academic research labs; as PARC's Tim Mott related to author Barry Katz: "We were basically building stuff. It was not 'Look at this important paper I published' as 'Look at this cool thing I made.'"[24]

In fulfilling their mission, PARC's management sought to embrace the creative agency of a Silicon Valley breed of researcher/developer, and together with them created their own kind of "office laboratory," seeing their working environment as an opportunity to support the kind of "thinking work" appropriate to their creative challenge. Rejecting a typical layout, PARC's Computer Science Laboratory conference room dispensed with the conventional large table, was lined with floor-to-ceiling white boards (then still a recent invention), and was home to the now infamous beanbags. It looked more like a lounge, befitting the image of the noncorporate work ethos of its employees. Scientists, artists, and others in PARC would drag the bags spontaneously to a relevant configuration, sitting cross-legged, reclined, or otherwise to listen and participate in the lab's weekly meetings. The beanbags ensured a contemplative posture; looking up or reclining back, the opposite of a typical conference-room posture. Not only did the beanbag give an air of informality due to its aesthetic and its unconventionality in a research laboratory, its placement was a strategic choice to support divergent thinking—and notably worked to prevent agitated researchers from jumping up and accosting presenters during more confrontational sessions.[25]

Sacco, the first commercially produced version of what we now call the beanbag, was designed as a "non-chair" only three years before the 1970 opening of PARC. Its designers Piero Gatti, Cesare Paolini, and Franco Teodoro sought to confront the propriety of "good bourgeois taste" and make a flexible piece of furnishing that morphed to fit the user's body.[26] This sack of polystyrene pellets looks nothing like a conference room chair—or any kind of chair for that matter—and arguably only becomes something like one when sat upon in a chair-like fashion. By its very design, it challenges norms and expectations, affording bodily postures and positions that a typical chair cannot. In the beanbag, one can sit back, nearly prone, looking up at the sky; drop into it, face down, knees bent, drawing in a sketchbook on the floor; two become an impromptu bed. Regardless of how it is engaged, the experience is quite different from that of a fixed piece of furniture designed with a narrow expectation of use. The beanbag greatly expands upon the postures that a "chair" can afford, engaging our informality through its specificity and alterity of form, material, and shape.

PARC's efforts—experimental, exploratory, unconventional, and very much "outside the box"—were both an outgrowth and enabler of the creative spirit of the early computing cultures of Silicon Valley. In embracing the alternative lifestyles of this particular ethos, it managed to produce a new form of laboratory culture and environment; its "distinctive combination of technology and place"[27] is among its many legacies.

Design Facts

By the time PARC scientists filled their conference room with beanbags, the thinking that designers do, or at least some forms of it, was well on its way to becoming, in proper noun case, "Design Thinking." David Kelley is a graduate of and longtime professor in the Stanford Joint Program in Design, cofounder of the internationally renowned interdisciplinary design firm IDEO and Stanford's Hasso-Plattner Institute of Design (more commonly known as the "d.school")— what Kelley calls the "ground zero" of design thinking. Kelley had the following to say about design thinking in a recent interview:

David Kelley (DK): All those years I said "You're experts at design methodology," nobody paid attention. They didn't take it as a new idea or a novel idea. They didn't believe it. For some reason, the words "design thinking" resonated with them.

Maria Camacho (MC):   Why do you think that the term "design methodology" didn't resonate?

DK: It sounded too much like other things. There's scientific methodology ... the word methodology has many other contexts. And the term design thinking, with the word "thinking," was just novel enough to attract attention. To put the words "design" and "thinking" together made both ideas new.[28]

The thinking that designers do takes many forms, and has, as we mentioned earlier, a rich history of explorations that have attempted to articulate what comprises it. The form of design thinking that "was just novel enough to attract attention" at Stanford's d.school is but one set of methods designers may use, but the concept has achieved vast amounts of attention since the creation of IDEO (1991) and the d.school (2004). In the interest of advancing creativity in the modern organization, Kelley and others have promoted their version of design thinking in workbooks and workshops as a step-by-step process to achieve creative results. Much of what they have folded into their methodology draws on the rich body of work on the thinking processes of designers, but has risked the very commodification and stabilization that one could argue its sources sought to resist. Bruce Nussbaum, once a faithful adopter of design thinking methods, reflects on its downfall:

I would add that the construction and framing of Design Thinking itself has become a key issue. Design Thinking originally offered the world of big business—which is defined by a culture of process efficiency—a whole new process that promised to deliver creativity. By packaging creativity within a process format, designers were able to expand their engagement, impact, and sales inside the corporate world. Companies were comfortable and welcoming to Design Thinking because it was packaged as a process.[29]

The commodification of design process isn't exactly new. With roots in the domestic engineering efforts of the 1920s[30] and the consumer engineering practices of the 1930s,[31] structured design processes helped industry by allowing for its generalization and exchange. Even the more radical human-centered approaches were co-opted by market pressures, and in many cases standardized into repeatable participatory and ethnographic design methods at the service of a growing consumer market.[32]

To see how easily these methods can become crude instruments, we can tease out what happens when even Latour's own "naïve observer" ethnographic approach—a method also commonly used by designers to study the ultimate users of their design products—itself falls prey to the microprocessing of facts. As an applied practice, designers obligate themselves to leverage their research into designs intended to be realized in some way for use in the world. In doing so, they must instrumentalize their observations, responding to each one in some material form. Let us imagine Latour as a design researcher charged with composing a new lab for these scientists. He observes a lab work process that is separated into one area that "contains various items of apparatus" where people "can be seen to be cutting, sewing, mixing, shaking, screwing, marking, and so on," and another that "contains only books, dictionaries, and papers" where people are "reading, writing, or typing," further dividing themselves into people who "write and engage in telephone conversations" and "those who type and dial telephone calls."[33] Design researcher Latour is "thus confronted with a strange tribe who spend the greatest part of their day coding, marking, altering, correcting, reading, and writing."[34] Armed with this knowledge, our fictional Latour might exclaim, "Clearly, these people need a cutting station and a mixing room, and these should be far separated from the reading room and writing center so that the two populations who do different kinds of work don't bother each other in the process. Plus! We need some kind of device that allows the people who dial telephone calls to pass them along to those who actually talk on the telephone, as these tasks must be separated for a reason."

None of Latour's observations (as ethnographer and theorist), by his own admission, reveal the thought processes of the scientists that yield their ultimate results;

they reveal important social and spatial contingencies that impact their thinking, but do these contingencies aid or hinder it? And what is the result of replicating them in a subsequent design?

The case of PARC's beanbag importantly raises a fundamental question of design. The object was recognized by those who purchased it as having a particular capacity to create an alternative to the conventional research laboratory. While they may not have been trained as designers, their understanding of the creative process and ethos of the lab led them to identify a designed artifact that they perceived as fulfilling a set of needs, and then ultimately structure their own workspace with it. PARC's deployment of the beanbag was based on a hypothesis that it would contribute positively and productively to workplace culture. Whether it did or didn't (an impossible to prove causality), PARC's remarkable creative output drew attention to the lab, and, as Stewart Brand valorized in the pages of *Rolling Stone*, was in large part attributed to the culture of the lab as much as to its resident expertise. Designers find source material in precedents, and with the tremendous press that the creative cultures of early Silicon Valley were receiving in the 1970s, exemplars like PARC became prime sources. As much as Latour and Woolgar describe the evolution of a statement from speculation to fact, so too do design speculations. Research is conducted, hypotheses are tested, experiments are executed, and through cycles of credit in the agonistic field of design publication, results are premiated. As the laboratory evolved from PARC to the Googleplex, the beanbag-populated workplace "featured as the object of debate" and served as "the focus of several published papers," becoming a key inscription device in the construction of design "facts."[35]

What we consider the "design fact" is not only the set of claims about what does or doesn't work for a particular situation, but the resulting design itself (whether proposed or materially realized), for all of these "statements" live on in the agonistic field of design. The liability for field observations, or any design research tool for that matter, conducted in the interest of human-centered design to be so crudely translated into simplistic design results can turn a revealing ethnography into a fact-generating apparatus. The limit is not inherent to design research, but to an unwillingness for researchers and designers to embrace the full gamut of microprocesses that comprise the observed activities. None of this is made any simpler for the designer if the humans at the center of their investigations are themselves unaware of, or unable to articulate the contingencies that influence their activities. Designers, in either of these cases, have little to go on, and can be seen to turn to already-established arguments, designs, or design research to influence their proposals. As will be introduced shortly, the results of this process, at least in the field of the creative knowledge worker, have advanced a design style that remarkably continues to reinforce an image of creativity that is already decades old.

A Laboratory Lifestyle

Architect and critic James Russell identifies a key trend of the knowledge economy in *Form Follows Fad: The Troubled Love Affair of Architectural Style and Management Ideal* (2000): "businesses began to recognize that white-collar work was more than pushing paper—it was pushing ideas. Companies needed more flexible and interactive ways of conveying and working with knowledge."[36] Citing the work of Studios Architecture as an important player in this transformation, Russell quotes principal Gene Rae: "in a knowledge economy, what you need is inside people's heads. People need to be encouraged to find out what they don't know and to share what they do know. We make workspace experiences where these things can happen."[37]

There was a particular fervor in the dot.com boom of the 1990s, and the startup culture that succeeded it, for continued references to the aesthetic qualities of the early Silicon Valley. "As the personal-computer industry exploded from tiny roots in dingy spaces, the garage paradigm took on the glow of creation myth. As these firms grew, they wanted to hold on to the fever of that nothing-to-lose startup era."[38] There is no shortage of commentary on what *Fast Company* magazine in 1993 identified as the workplace ethos behind these designs: "neo-leisure," the work hard/play hard mentality of the evolving knowledge worker workforce.[39] Numerous major mass-market magazines and news sources have run laudatory articles-as-PR—too many to count. There have been critical inquiries[40] and there have been scathing exposés.[41]

Creativity and innovation, at the moment, appear to work best in some combination of scattered furnishings with a wide range of forms, material, and accommodation, combined with eye-catching "centerpiece" elements that are clearly not "of" the workplace—slides, game tables, gondolas, vintage (or new but intentionally "distressed") furnishings. Various design researchers point to different and regularly conflicting "facts" about worker productivity, but ultimately, we find what Thomas Davenport identifies in the ominously titled *Thinking for a Living: How to Get Better Performances and Results from Knowledge Workers* (2005): "At least there is no evidence that anyone ever took a job, stayed at a job, or worked more productively because of foosball, pool, or ping-pong tables, cappuccino bars, office concierges, hearths, conversation pits, quiet rooms, lactation rooms, creativity rooms, relaxation rooms, nap rooms, etc."[42]

Nevertheless, according to James Russell, Donald Albrecht, and Chrysanthe Broikos, since the days of PARC,

just-in-time spaces, hoteling, host-desking, coffee-shops, game rooms, commons' and hives became the lingo of the day, all seeking to identify their hosts as markers of innovation, creative
thinking, collaboration, and work-as-play. Architects and interior designers use spatial drama
and architectural materials to signal the youthful, no-holds-barred nature of the workplace, as
an advertisement of corporate values, and even as decoration meant to express the [perceived]
nonhierarchical nature of the space. In this topsy-turvy work world, the fitness center or coffee
bar gets the most lavish architectural attention.[43]

As designers and their clients seek greater distinction for their work, in particular
through signifiers of creative aspiration, we have reached a point where even a "forest" of tropical plants (Google Zurich), Ferris wheel (Acuity, in Sheboygan, Wisconsin),
"cloud room"(VMware), playground slide and ball pit (iSelect in Melbourne, Australia),
or literal tree-house, tree included (Davison, in Pittsburgh, Pennsylvania), no longer
surprise with their deviance from the image of the staid laboratory. The "neo-leisure"
design theme is intended to cater to the particular lifestyle of the modern knowledge
worker: that twenty-first century descendent of the early Silicon Valley computer bum.

In 2014, the furnishings company Steelcase, known for their trend-forming exploratory design research conducted in collaboration with renowned design consultants,
conceived and realized a full-scale fully-functioning prototype "Innovation Center
Typology," a "Palette of Place, Posture, and Presence."[44] The litany of design facts in
the project's narrative, such as "Innovation requires a connection between sociology
and technology" or "Creative, generative collaboration happens in small groups," are
carried through to the design, whose architectural diagram reads like a theme-park
map, with a "front porch," "innovation theater," "neighborhood café," and multiple
"enclaves" called out as if attractions in Workplaceland. Even the less photogenic areas
of the showcase interiors are filled with many variants of things; the "hot" cubicle—an
unassigned furnishing artifact that separates individuals from their otherwise open,
flexible, reconfigurable environment; the unenclosed or translucent-walled "strategy
room" (once known as a conference room); and seating artifacts that, as corporate-
managed, scientifically researched evolutions of the beanbag, present the user with a
great variety of possible postures for thinking.

As a design "fact," the equation "treehouse + enclave + slide = creativity and innovation" has reached that point of "naturalization," "whereby all traces of production are
made extremely difficult to detect."[45] We don't intend to make the claim that one can't
think creatively in these spaces. Sometimes a treehouse or a "front porch" would be
perfectly productive for some kinds of work. Our point is that these designs, although
intended to support creative thinking, are prone to miss the nuances that comprise
their stakeholder's own thinking processes and creativity. They offer a set of environmental contingencies that may have more to do with the design professions and their

concerns than with their user's particular "thinking needs." Worse, in the guise of creative openness, they may serve to mask the traditional corporate power dynamics that may not be compatible with the apparent creative freedom. Will Alsop's design for the Blizard Building, at Queen Mary University of London exemplifies this point for the laboratory. The "mushroom pod," together with the "cloud pod" and "spikey pod" seen in the background, and the "molecule pod" unseen in figure 2.3, are meeting, lounge, and exhibit spaces that hover over the subterranean laboratories below.

Heinrich Schwartz highlights the point that invisible power structures not only remain in existence with these organizational changes, but also become perhaps more insidious due to their invisibility. He argues that these new power dynamics may no longer be based on space ownership, but rather on the "mobility and ownership of technology, or 'techno-territory,'" a "re-drawing of hierarchy rather than its disappearance."[46] In one of the sites he investigated, he describes how "the club," a casually furnished, open plan, unassigned work area "had in fact become a rather exclusive area. … Intended to break down barriers and reduce hierarchies within the company, the supposed flatter design appeared to introduce new spatial boundaries, most notably between mobiles and non-mobiles and between regular and higher-ranking employees."[47] The spatially open, aesthetically different, and more relaxed office environment thus maintains a symbolic value that is more complex than it may appear. This is not to say that its style belies its function. Quite the opposite: its style *is* its function.

The challenge with the lack of awareness of the constructedness of design facts—as with the scientific facts observed by Latour and Woolgar—is that they appear "out there," autonomous from the very meaningful processes that led to their articulation. These facts take on the qualities of being unassailable and natural. "The beanbag *is* a prompt for thinking out of the box," rather than "In the 1970s, at this unique center in Palo Alto called PARC, they used the recently available beanbags as their primary meeting room chair." The latter statement is certainly less of an elevator pitch, but acknowledges some of the many factors that connected the beanbag to the knowledge worker industry. If we imagine a countervailing scenario, the beanbags at PARC may have been impediments to the creative work that the scientists were producing. As we recounted earlier, the most specific description of their function that we have come across (most accounts merely describe their presence) is of hindering altercations rather than fostering creativity. But this is not the image of the beanbag that we carry today.

The reductive applications of design research and design thinking to construct design facts reifies the equation of the beanbag with exploratory thinking. This is done not only in the stage of primary fieldwork or case study research, but also in that agonistic field of design, where academic and professional critiques, discourses, and

Figure 2.3
The interior of the Blizard Building at Queen Mary University of London, designed by Will Alsop (2005). Photograph by Morley von Sternberg. Courtesy of Will Alsop/aLL Design.

vetting premiate certain products, realized or proposed, over others. As with the scientists at Salk, this has the essential function of removing evidence of the constructedness of these products and their assumptions. What users are left with are, in effect, symbols of the prompts that may have, in other contexts or places or times, have had a positive effect on the creative process of the knowledge worker, rather than the prompts themselves.

As these facts continue to circulate and build on one another (whether it's the super-table or slide or treehouse or gondola that is now equated with the kind of thinking valued by today's corporations), we question what impact these symbols have as supports for any form of thinking beyond that which is codified by the brand image they are intended to reinforce. We've seen how many of the ostensibly "flexible" or "open" or "social" lifestyle designs of the contemporary knowledge industry lab have served to mask traditional power structures and dynamics. What, by extension, might be happening to the results of the thinking processes that are mediated—constructed—in the context of these same designs?

What, also, might the designer do in response to this critique? Can one avoid the agonistic microprocesses that construct design facts? Earlier, we argued that a primary contributor to an organization's or designer's reliance on and repetition of design facts is the opacity of the thinking processes for which they are designing, but equally opaque is the thinking of designers *themselves*. An acute awareness of a designed object's constructedness—a recuperation of the contingencies and "traces of production"—might be a necessary function of a designer's thinking process. This necessarily challenges the mythos surrounding a designer's "creativity," their ability to "invent" successful designs through a process that remains ineluctably idiosyncratic. Yet, at its core, it requires only the effort to engage a designer's thought process, not with a scientific analysis, but with a reflexivity that questions the "stability" of their design concepts. Reflexivity is thus a way of reminding the ~~reader~~ designer that *all* ~~texts~~ designs are stories. This applies as much to the facts of our ~~scientists~~ designers as to the fictions "through which" we display their work. The story-like quality of ~~texts~~ designs denotes the essential uncertainty.[48]

Notes

1. Richard Florida, *The Rise of the Creative Class, Revisited* (New York: Basic Books, 2012).

2. Bruno Latour and Steve Woolgar, *Laboratory Life: The Construction of Scientific Facts* (Princeton, NJ: Princeton University Press, 1986), 176.

3. Ibid., 168.

4. Ibid., 169.

5. Ibid., 171.

6. Ibid., 176.

7. Ibid., 174.

8. Ibid., 179.

9. See especially Herbert A. Simon, *Administrative Behavior: A Study of Decision-Making Processes in Administrative Organization* (New York: Macmillan Co., 1947); Herbert A. Simon, *The Sciences of the Artificial* (Cambridge, MA: MIT Press, 1969); Edward de Bono, *The Use of Lateral Thinking* (New York: Basic Books, 1968); Seymour Papert, "Teaching Children Thinking," in *Proceedings of IFIPS World Congress on Computers and Education* (Amsterdam: North-Holland, 1972).

10. Pierluigi Serriano, *The Creative Architect: Inside the Great Midcentury Personality Study* (New York: The Monacelli Press, 2016).

11. John Markoff, *What the Dormouse Said: How the Sixties Counter-Culture Shaped the Personal Computer Industry* (London: Penguin Books, 2005), 28.

12. Ton Otto and Rachel Charlotte Smith, "Design Anthropology: A Distinct Style of Knowing," in *Design Anthropology: Theory and Practice*, ed. Wendy Gunn, Ton Otto, and Rachel Charlotte Smith (London: Bloomsbury Academic, 2013).

13. Arthur Pulos, *The American Design Adventure, 1940–1975* (Cambridge, MA: MIT Press, 1988), 185.

14. Jennifer Kaufmann-Buhler, "From the Open Plan to the Cubicle: The Real and Imagined Transformation of American Office Design and Office Work, 1945–1999" (PhD thesis, University of Wisconsin-Madison, 2013), 27; James S. Russell, "Form Follows Fad: The Troubled Love Affair of Architectural Style and Management Ideal," in *On the Job: Design and the American Office*, ed. Donald Albrecht and Chrysanthe B. Broikos (Princeton, NJ: Princeton Architectural Press, 2000), 59.

15. Fritz Machlup, *The Production and Distribution of Knowledge in the United States* (Princeton, NJ: Princeton University Press, 1962).

16. Kaufmann-Buhler, "From the Open Plan to the Cubicle," 27–28.

17. Ibid., 29.

18. Ibid., 27.

19. Michael A. Hiltzik, *Dealers of Lightning: Xerox PARC and the Dawn of the Computer Age* (New York: Harper, 2000), 159.

20. Stewart Brand, "SPACEWAR: Fanatic Life and Symbolic Death among the Computer Bums," *Rolling Stone*, December 7, 1972.

21. For an engaging introduction to Bell Labs history, see Jon Gertner, *The Idea Factory: Bell Labs and the Great Age of American Innovation* (London: Penguin Books, 2012). Its historical narrative feeds directly into Hiltzik's *Dealers of Lightning*.

22. Andrew Ross, "Jobs in Candyland—An Introduction," in *No-Collar: The Humane Workplace and Its Hidden Costs* (New York: Basic Books, 2003), 1–20.

23. Barry M. Katz, "Research and Development," in *Make It New: The History of Silicon Valley Design* (Cambridge, MA: MIT Press, 2015).

24. Ibid.

25. *Adele Goldberg: Bean Bags and PARC Culture*, online video, http://www.computerhistory.org/revolution/input-output/14/348/2300 (accessed May 20, 2017).

26. "Sacco: Piero Gatti, Cesare Paolini, Franco Tedoro," *Vitra Design Museum*, https://www.design-museum.de/en/collection/100-masterpieces/detailseiten/sacco-gatti-paolini-teodoro.html (accessed June 6, 2016).

27. Nathan Ensmenger, "Beards, Sandals, and Other Signs of Rugged Individualism: Masculine Culture within the Computing Professions," *Osiris* 30 (2015): 38–65.

28. Maria Camacho, "In Conversation: David Kelley: From Design to Design Thinking at Stanford and IDEO," *She Ji: The Journal of Design, Economics, and Innovation* 2 (Spring 2016), http://dx.doi.org/10.1016/j.sheji.2016.01.009 (accessed June 6, 2016).

29. Bruce Nussbaum, "Design Thinking Is a Failed Experiment. So What's Next?," April 5, 2011, https://www.fastcodesign.com/1663558/design-thinking-is-a-failed-experiment-so-whats-next (accessed June 6, 2016).

30. Christine Frederick, *The New Housekeeping: Efficiency Studies in Home Management* (Garden City, NY: Doubleday, Page & Company, 1918).

31. Roy Sheldon and Egmont Arens, *Consumer Engineering: A New Technique for Prosperity* (New York: Arno Press, 1932).

32. David Raizman, *History of Modern Design: Graphics and Products since the Industrial Revolution* (London: Laurence King Publishing, 2003); Wendy Gunn, Ton Otto, and Rachel Charlotte Smith, eds., *Design Anthropology: Theory and Practice* (London: Bloomsbury Academic, 2013).

33. Latour and Woolgar, *Laboratory Life*, 45.

34. Ibid., 49.

35. These are the phrases used by Latour and Woolgar in *Laboratory Life* to describe the reification of actions and inscription devices. Latour and Woolgar, *Laboratory Life*, 66.

36. Russell, "Form Follows Fad," 60.

37. Ibid., 70.

38. Ibid., 69.

39. Fast Company, "A Spy in the House of Work: Report #1: Neo-Leisure, the Dirty Little Secret behind the 65 Hour Workweek," October 31, 1993, https://www.fastcompany.com/54955/spy-house-work-0 (accessed May 20, 2017).

40. Ross, *No-Collar*; Gina Neff, *Venture Labor: Work and the Burden of Risk in Innovative Industries* (Cambridge, MA: MIT Press, 2012).

41. One particularly relevant example is Dan Lyons, *Disrupted: My Misadventure in the Start-Up Bubble* (New York: Hachette Books, 2016).

42. Thomas Davenport, *Thinking for a Living: How to Get Better Performance and Results from Knowledge Workers* (Boston: Harvard Business School Press, 2005), 169.

43. Russell, "Form Follows Fad," 70.

44. Steelcase, "The New I.Q," *360 Magazine*, no. 7, https://www.steelcase.com/eu-en/research/articles/topics/innovation/the-new-i-q/ (accessed May 20, 2017).

45. Latour and Woolgar, *Laboratory Life*, 176.

46. Heinrich Schwartz, "Techno-Territories: The Spatial, Technological and Social Reorganization of Office Work" (PhD diss., Massachusetts Institute of Technology, 2002), 2 and 189.

47. Ibid., 195.

48. Latour and Woolgar, *Laboratory Life*, 284.

3   The Good Experiment: How Architecture Matters for Graphene Research

Albena Yaneva and Stelios Zavos

Space Matters for Science

The founders of science and technology studies (STS) Bruno Latour, Steve Woolgar, Michael Lynch, and Karin Knorr-Cetina pioneered meticulous explorations of *science in the making* to understand the process of the fabrication of scientific truth and facts; the cognitive and social dimensions of scientific experimentation and visualization; and the material operations that accompany scientific work.[1] From the 1980s, Peter Galison, Steven Shapin, Sven Dierig, Jens Lachmund, and Peter Mendelsohn have engaged in addressing questions of how space, locality, urban infrastructure, and city development matter in the production of scientific knowledge.[2] Taking inspiration from urban studies and architecture, they have focused attention on the importance of space to the credibility of scientific claims. They have also tackled the question of how the urban infrastructure and the architecture of various scientific buildings and laboratories, as socio-spatial settings, affect the production of knowledge and work patterns, and thus challenge the cognitive authority of science. These recent dialogues between the fields of science studies and architecture have made us rethink architecture's role in the shaping of scientific cultures and identities,[3] the situatedness of scientific activities,[4] the importance of space for both the production of scientific knowledge and the credibility of scientific claims,[5] and the complex nexus of knowledge and space.

At the same time, STS scholars have expanded their methods to engineering, design, technological innovation, medicine, economics, and the arts, by following the actors in their routine practices, accounting for their actions and transactions in complex spatial settings, and unpacking the materialization of the successive operations they perform. Following this expansion, architecture has also received the attention of anthropologists trained in science studies, such as Michel Callon, Sophie Houdart, Yanni Loukissas, and Albena Yaneva who have offered an alternative pragmatist understanding of architecture-making, very different from the one bestowed by the critical theory that dominates architectural discourse.[6] Inspired by the pioneering work of Dana Cuff, these studies traced architecture in the making.[7] Consequently, on the one hand, the interest of the STS community in issues of architectural and urban design has increased.[8] On the other hand, architectural theorists have also started referring to the epistemological frameworks of science studies and have begun borrowing concepts and methodological insights.[9]

In spite of the growing cross-fertilization of the two fields, whenever scholars of architecture deal with scientific buildings, they often ignore the long tradition of STS that studies what happens *in* labs, namely the socio-material complexity of the practices of dwelling in labs. As a result, *lab life* remains entirely forgotten or considered as insignificant to the understanding of how architecture works. Similarly, whenever STS scholars tackle the architecture of science, they rarely discuss the specific architectural features of labs (e.g., location, site, facade design, visual language, design constraints, the negotiations of designers and client-users, and the specific materials rearrangements). As a consequence, the design and planning processes behind scientific buildings were rarely accounted for with the exception of the work of Thomas Gieryn, Peter Galison, and Emily Thompson.[10] Nevertheless, the practices of science labs and of architecture studios bear an astonishing resemblance to each other; both types of practitioners deal with trials, produce different scenarios and options, present results from experiments with materials and shapes, engage in measurements with models, simulations, and calculations, meet clients, funders and potential users, and take into account public reactions.

In our attempt to develop further this dialogue between the disciplines, we draw on the methods of the anthropology of architecture and science studies to trace the exchange between the designers and the scientists involved in the making of the National Graphene Institute (NGI) in Manchester, England. We provide glimpses into the inner workings of their labs and trace the unfolding dynamics of experimentation that preceded the design. If, for the seminal studies on the architecture of science, the key question was to explore both the relationship between the buildings and the shaping of the identities of scientists and their fields, and also the identities of the architects who design them, for us our key concern is different. Instead, we will ask what scientific architecture can tell us about the changing nature of scientific practice today, and more importantly, about the changing larger networks of scientific production (the partnerships with industry, the city, and their funding structure). In other words, our question is: How do the dynamics of the new ecology of the science industry inform us about the shifting practices of the architects and the working dynamics of their "architectural labs"? Or, how does science lead to new tactics, tools, and techniques in the practices of architects and building designers? Instead of navigating between architectural determinism from one side (architecture determines the science conducted inside) and architectural indifference on the other (architecture is irrelevant to the science contained within its walls), we will trace how a very specific scientific breakthrough—the isolation of graphene—acted as a complex machine that reconfig-ured both scientific and architectural practices.

Designing the "Home of Graphene"

In 2010 the Nobel Prize in Physics was awarded to two physicists from the University of Manchester—Andre Geim and Kostya Novoselov—for the isolation of graphene, the first man-made two-dimensional material. Five years later, in 2015, following the ambi-tion of the university to capitalize on the work done by its Nobel Prize winners, a new building was erected on the campus: the National Graphene Institute. Located at the main university campus on Booth Street West, the NGI draws the attention of visitors—a five-story building of around 7,825 square meters (approximately 84,230 square feet) with a distinctive black silhouette. Distinctive as it is, it breaks the monotonous patterns of gray and red brick buildings on the campus; its unique shape and sizable volume draw media attention, and the building enjoys a growing public interest as a showcase of cutting-edge science. Just a few minutes' walk from the School of Architecture build-ing, we—who are colleagues at the university—have often passed alongside it, contem-plating the building site and awaiting impatiently for it to take shape.

However, to understand the building, we first need to understand graphene, the foil of nanotubes, whose properties have been known for decades, but the extraction of which as a single mono-layer with distinctive electrical properties and strength occurred only in 2004. This marked the start of the process of isolating the material. The applications of graphene have yet to be explored, and the new building will serve as an incubator dedicated to its development by bringing together academics and com-mercial partners under one roof. Its design and state-of-the-art facilities are meant to contribute to the UK's role at the forefront of the commercialization of graphene. The facilities will allow scientists and engineers to further explore how graphene interacts with other materials, and to develop prototypes that could potentially enter into full production. Currently, 150 researchers work directly on graphene at the NGI, and more buildings dedicated to graphene research are under construction on the campus of the University of Manchester.[11]

The chapter draws on in-depth interviews with scientists, architects, university man-agers, building managers, and cleanroom technicians engaged in the recent process

Figure 3.1
The National Graphene Institute, Manchester. Photograph by the authors.

of the design and construction of the NGI building. Our ambition was to shed light simultaneously on the life of two "labs"—that of the architects as well as that of the scientists. We witnessed the inner life of the NGI, those aspects of the design experienced by residents but hidden to the public, in its many different labs and communication spaces. We observed the active collaboration of scientists energetically shaping the building through experiments, discussions, and negotiations. On the other hand, we also gained unique insights into the "architectural lab" of the London-based firm Jestico + Whiles, as they engaged in BIM experimentation and design discussions.

Additionally, we conducted ethnographic observations in different spaces in the building. But due to the numerous restrictions (as isolated and protected environments set barriers for our access), we were not allowed to randomly stroll and explore the spaces. So, in order to gain a better understanding of how the building works for various groups of "dwellers," we took ethnographic walks with the interviewees, asking them to recreate their daily trajectories and the specific ways of engaging with the different features of the building. This allowed us to gain insights into the different practices and routines of dwelling that form the core of laboratory life at the NGI and contribute to a "lifestyle science" related to graphene research. During the ethnographic walks, we stopped many times, questioned their attachments to the building, took photographs, and explored the different material arrangements, equipment settings, inscription techniques, and the various design features of the building that mattered for our NGI dwellers. Following the rhythm of scientific dwelling in its course, the ethnographic walks advanced a different understanding of the nature of scientific buildings and shed light on how architecture matters for the scientific practices of nanotechnology research. In addition, we visited the cleanrooms, and wandered the viewing corridor many times trying to imagine what a random visitor would see from outside. We sat in quiet labs, and witnessed the buzz of the busy Fridays, when more than a hundred people assemble at the building for the famous graphene seminars. We documented the building and captured its working rhythm in 2017, two years after it was built and was entering a steady-state phase of functioning.

While analyzing the work and lifestyles of the two labs, we will zoom in on the nitty-gritty reality of the process of experimentation, the material tests, the techniques, the patterns of collaboration, as well the modalities of social exchange. Yet, before entering the NGI to witness graphene research in action, it is worthwhile exploring some prior examples of successful and less successful collaborations between these two types of practitioners and their labs, traversing quickly through the changing landscape of lab designs in the last decades.

The Changing Formulae of Science-Architecture Partnerships

From the 1960s onward, signature architects showed more interest in designing scientific buildings. The dialogue between architects and scientists became an important factor for the success of scientific labs. Yet, very often, poor understanding of the nature of experimental practices in laboratories resulted in buildings that were deeply disliked by the scientists. The Richards Medical Research Laboratories (1965) designed by Louis Kahn and the Ray and Maria Stata Center (2004) by Frank Gehry are notorious in that respect. Kahn's lab buildings in the 1960s foreshadowed in many ways today's laboratory design. The Richards Medical Research Laboratories at the University of Pennsylvania, his first scientific building, was greatly admired in architecture literature for its imposing presence and imaginative presentation of space and structure. However, the scientists complained about exposed pipes that collected dust, the lack of wall space for refrigerators, and about sunlight penetrating the building, melting ice in the buckets, and spoiling experiments. The building *as experienced in mundane lab life* routines appeared to hamper rather than facilitate research. The contrast between the building's beautiful shell and the inner working of the labs remain striking. Not surprisingly, the scientists were rarely included in design discussions.

However, Kahn's second lab building, The Salk Institute for Biological Studies, in La Jolla, California (1965), was operationally more successful. Working in close collaboration during the design, Salk and Kahn envisioned the building together. Learning his lessons from the Richards lab, Kahn built a place in which scientists felt comfortable working and, as a result, the building won plaudits for being functional: it accommodated flexible lab facilities, essential to the fast-changing world of science, allowing ease of updating the mechanical equipment. The open labs encouraged students and postdocs to socialize. Flooding the laboratories with indirect daylight and producing an open and airy work environment, Kahn's building provided a welcoming and inspiring environment for scientific research and formed the first example of a successful "signature" architect lab. Other examples followed. Among them, Payette Associates, and Venturi, Rauch and Scott Brown's Lewis Thomas Laboratory in Princeton (1983) stands out. Developed around the "generic" laboratory model, it followed the large open lab design pioneered by the Salk, and emphasized the importance of discussion for the successful realization of molecular biology by including a number of generous circulation and breakout spaces. The contradictory desires for social exchange and the seclusion and separation between scientists, because of the specific environments of their work, were overcome efficiently through the design of the Lewis Thomas Laboratory. Since then, a new generation of scientific buildings have explored the facilitation of interactive behavior further, and promoted social exchange and collaboration among researchers from different disciplines through the open lab model. In these buildings, the atrium became more important than the laboratories themselves.[12] The most recent trend, the subject of this book, shows a new generation of scientific buildings that emerge as megastructures or complexes of buildings that form entire self-sustained quasi-urban structures where scientists live and work, and where the boundaries of work and leisure dissolve.

This brief history of several decades of architectural interest in laboratory design forms the background of our case study, but the collaboration between scientists and designers here takes a specific form and has its own particular emphasis and effects. In particular, a unique partnership is forged between a Nobel Prize laureate, Sir Kostya Novoselov, and the architect Tony Ling, director of the London-based architectural firm of Jestico + Whiles, which resulted in a building that is one of the new generation of sustainable, high-tech science laboratories aimed at a collaborative research culture.

The National Graphene Institute: Kostya in Search of the Right Design

Reflecting on the architecture of the building, the different issues and compromises, the lead scientist Kostya summarized it as "a continuous fight between Tony and me; I was trying to reduce the architectural features, he was trying to enhance them. So, what you see is *basically the result of this battle*, but we tried to maximize the space and make it as flexible for the future as possible."[13] Amazingly, the "battle" between the reduction and enhancement of the architectural features has resulted in a building the scientists at the NGI are happy with. Yet, for Kostya, who was often referred to by fellow scientists as "the true architect of this building," the NGI building is quite logical. Embracing design language, he evoked different constraints at the start of the design process: first, the need of large cleanrooms, and second, the need to build the NGI at a walking distance from key departments, like Physics and Chemistry. He explains that he "wanted the building to be very universal in a way that, for example, all the architectural features can be essentially converted at a certain moment, to be useful space."[14] We probed this concept of universal and adaptable space further as we walked around the building with Kostya. In response, he showed us where he wanted some anchor bolts to be installed in the free space behind the main glass facade, as well as in the atrium of the building, so that metal beams could be placed in the future if needed. This modification would create additional floor area to be utilized as office space. As we walked, we also learned that some design features were incidental—like the roof terrace and garden on the top of the building—and he told us about the many different ways these features can be used advantageously. Kostya is happy and proud of this building; nevertheless, he did not tell us what the building *is*, rather how the building will grow. He thinks constantly about the opportunities to add, increase, expand, and maximize its efficiency.

At one point, we stopped on the roof level to contemplate the facade where Kostya recalled the process of facade design as being "the worst point of discussion": the architects came to him with the idea of a veil that he thought was good but too expensive. However, at that moment of doubt, Kostya felt it was the scientists' turn "to give them [the architects] something back because they gave a lot to us, but of course it was very difficult to choose something for the image."[15] This process of negotiation required a lot of experimentation together. Tony and his team produced and studied hundreds of different options of what the building should represent. It took a lot of time: long hours of discussions for the architects and the scientists to explore different patterns. This was a moment of intense experimentation, a moment when the visuals proliferated, and

Figure 3.2a, b
The facade of the NGI; close-up. Photographs by the authors.

hundreds of renderings traveled hectically between London and Manchester.[16] Kostya remembers:

At a certain moment, they came with a sort of image, like in this movie, *The Matrix*, where the numbers kind of fall down and then start forming a pattern. So, the architects came with those numbers and then I said "ok, why won't we just put some formulae" and then we had very, very good hand-written formulae and then they converted and digitized them in such a way that you wouldn't be able to read them, but you'd be able to see that those are formulae.[17]

Both Tony and Kostya agreed that the formulae should be very subtle. In our walking discussions, the authors and Kostya dwelt a great deal on these almost undetectable graphene equations inscribed in perforated metal. However, we had passed along this building so many times, and to our embarrassment, had never noticed the formulae. Yet, now aware of their conceptual generation and presence in the building's facade, they make sense. At one point, the university administration requested that the formulae be legible, and the architects and scientists tried several ways—hand-written, typed—then copied all the formulas from the allocators of graphene there and put them on the facade. They also added "legitimate mathematical jokes," as Kostya calls them, so correct that they can be directly copied to textbooks. Thus, the facade was a compromise. The architects considered an opaque black screen, but this would have been too weighty. Another option was to have large hexagons, but Kostya did not want a very explicit reference to graphene to be placed on the facade; they only kept the little hexagonal perforations. We recollected the intense discussions and negotiations between the architects and the scientists around these different facade options. The Matrix-movie moment yielded a very interesting exchange: while Tony was learning to read and decipher the different formulae of graphene, grasping simultaneously what graphene is, and how it is translated into equations; Kostya was learning how to read the different shades of black and gray on the architectural rendering and how to imagine the facade on the basis of the visuals produced in Tony's lab. Staring at the facade formulae now, looking to find Kostya's "jokes," puts us in the mood of a worldly view of science, science that never happens in a "double-click" moment of invention, but is rather an endeavor that takes numerous reiterations with mistakes in search of the right answer—that is, science in the world.

The Other Lab: Tony Ling in Search of the "Right Formula"

At the same time, in the other lab in London, architect Tony Ling and his team experienced the challenges of designing the NGI differently from the scientists. Tony listened to the stories of many graphene researchers to understand how they work, to grasp their expectations and the nature of their practices. For him, the texture of spaces and surfaces should reflect the funding realities and the discussions with the scientists; the changing needs and preferences of different research teams; the vision of the lead scientist, Kostya, and, the very complex texture of graphene and its projected development. As a scientific building, the NGI followed an inside-out approach wherein the form grew from the internal arrangement of the programmatic parts and their gradual resolution. While describing the exciting collaboration he had with Kostya in designing the building, Tony emphasized how exceptional it was to have a client involved in the process from day one. He recalls Kostya spending, literally, hundreds of hours working with the Jestico + Whiles at every step, assisting in determining the design, development, and the relationships between these phases.

Although Tony has experience in designing nanotechnology buildings (for the University of Southampton and the University of Sydney), he finds that

the main difference between the NGI and other nanotechnology-related buildings, is the fact that they are generally part of the Physics department ... whereas in the NGI, the sole purpose of it is to explore the properties of graphene, and also, the NGI has a policy of invitation of industry partners, who are given space within the building, to work with the academics at the university, to mutually benefit from the research.[18]

Thus, for Tony, graphene and the properties of graphene were at the core of the design concept. The architect tried to understand as much as possible the nature of graphene before his team started work on the building, but, as he shares honestly with us "the actual, very detailed properties and mechanics of working with graphene are beyond our grasp," and the architects' main job was to understand what kind of design environments the scientists working on graphene will need.[19]

The very first design decision that the design team, the structural engineers and Kostya had to take was where to locate the cleanroom. This is the most important component of the building as it is where the nanoscale research into graphene takes place. The technical performance of the cleanroom is paramount for the building's success. The NGI building contains two cleanrooms: the main cleanroom takes over the lower ground level for minimal vibration and is connected by a cleanroom lift—the only one in England—to a second cleanroom on the third level. One of the key initial decisions Tony and Kostya made was to locate the cleanroom in the basement, four meters below ground level because the material is to be studied in such a small scale, any kind of vibration could disturb the experiments. This started out as the driving force behind the rest of design; the building followed the cleanroom and its form naturally grew out from the program.

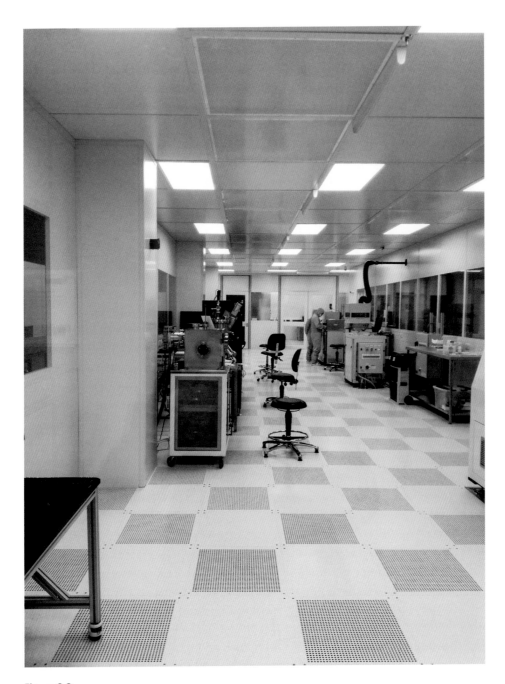

Figure 3.3
The large cleanroom at the NGI. Photograph by the authors.

The next challenge was the discrepancy between the expectation for a 5,000-square meter (54,000-square foot) building, according to the university's feasibility report, and a much bigger building, calculated on the basis of the scientists' needs. For six weeks, the designers engaged in analyzing the net to gross floor ratios and cost per square meter of similar buildings with cleanrooms that had been recently completed, producing different options and comparative scenarios. Finally, the tests proved that a larger building was necessary (84,228 square foot or 7,825 square meters). The multi-headed body of stakeholders (project managers, the University of Manchester, government funding agents) had to readjust the budget accordingly. The initial funding for the building came from the UK government, and the European Regional Development Fund (ERDF) granted the extra money to construct the building. As Tony recalls: "it was quite an intensive exercise to eventually challenge that the building was to be bigger."[20] While architects probed different scenarios, tested options and refined their working techniques, mobilized new tools and engaged in experimentation, the scientists fine-tuned their expectations and the funders increased the budget.

Following these two challenges, the architects had to engage in more discussions with Kostya to understand what the specific requirements of the NGI were. Due to the complexity of the building it took weeks of iterative interviews with the scientists; along with repeated attempts at schematic plans and three-dimensional models, so that the architects could truly understand what the scientists, as clients, wanted. Tony and his team engaged in a process of generating "experimental layouts of notional ideas," which they presented to Kostya, to say "is this what you mean?" Numerous discussions followed from which both the architects and the scientists learned from one another. As a result of experiments and trials in the design lab, these images also served as a way of communicating the expectations of the scientists. They allowed designers to learn from the user-clients and at the same time helped clarify structural, mechanical/electrical, and IT issues.

The next stage that required an active exchange between design and scientific know-how was the process of organizing the building according to the functions, an approach that the architects term as "inside-out." During this process architects made numerous models, both virtual and physical, of the whole building, as well as of the different components of the building. They engaged in three-dimensional modeling, and even produced a 1:50 model of the rear facade of the building to see how it really fits together. The models illustrated different options and took into account cost and program factors, well as planning and university contexts. They also showcased the inside-out approach: how starting from a number of "knowns" the building grew around the program and took a distinctive shape that followed its internal organization. The

Figure 3.4a, b
Experimental 3D models and drawings developed by the architects Jestico + Whiles. Courtesy of
Jestico + Whiles.

knowledge the architects gained in this project is now being transferred to the design of the Cavendish Lab in Cambridge. While the design process of the NGI implied fine-tuning the tools and techniques of both architects and scientists, it also impacted the way the building mattered in the daily practices of graphene research. In order to witness the "inside-out" approach, we should go inside the building. Let us take a quick walk through the building in use.

A Walk in "Graphene Land"

On our first visit to the NGI, we found ourselves led straight to the top floor, where the seminar room opens onto a large rooftop terrace. In addition to the key scientist, Kostya and the architect, Tony, we interviewed building managers (John, Fran, and Richard), the scientific director (Vladimir), and scientists (Mark, Andrey, Sarah, Rahul). The ethnographic walks with a few scientists often began here, on the top floor, over coffee with Kostya, with Mark showing us the garden, and John introducing us to a happy group of young researchers and postdocs ("the core graphene team") having lunch on the terrace in the shy Manchester July sunshine. We were introduced to the cleanrooms, and visited them many times to witness dark-blue-gowned technicians and light-blue-gowned researchers exacting minute procedures on graphene flakes.

There are eighteen different labs in the NGI, in which we witnessed mixtures of different disciplines: a laser lab, an optical lab for electrical measurements, and metrology and chemical laboratories. The building accommodates the many different scenarios these laboratories pose to risk management, their different hazards, diverse approaches, myriad technologies, inspection procedures, multiple outcomes, and different algorithms. The combination of physicists, chemists, and material scientists working together sets challenges for the spatial design.

One aspect that the scientists were very much concerned with was the "gray spaces": special seven-foot- (two-meter-) wide corridors, adjacent to most of the labs, where tools and equipment are stored, and air and electrical systems and gas piping are installed. If the nature of the lab is to be changed in the future, the "gray spaces" enable this flexibility to be done quickly and easily. Each lab has its own specific qualities. For instance, some of the labs are located along the exterior of the building, maximizing natural views and light and others are secluded or shielded. During one of our visits, we surprised a postdoc taking a nap in the Faraday chamber (fully isolated from electromagnetic waves). Embarrassed, he explained that he had had a long night in the lab and did not get to sleep at all; and at 4 pm he was still there waiting for the experiment to finish. In another lab, we met a happy postdoc listening to loud music; again, we

Figure 3.5
Atrium and roof terrace on the top floor. Photograph by the authors.

Figure 3.6
Lab spaces at the NGI. Photograph by the authors.

surprised him, yet, he responded with a wide smile to our visit. There are also open labs connected to the offices of senior scientists. In addition to labs, we also witness a large number of "breakout" rooms with furniture designed to encourage a convivial ambiance for discussion.

During these walks and interviews we noticed, on the one hand, that the NGI successfully maintains the dual requirement for isolated, clean, protected environments away from vibration and electromagnetic interferences, where no noises from the external world of the lab should be allowed to inhibit results. We also noticed, on the other hand, the importance of highly interactive and collaborative spaces for research dialogue and developing new partnerships. This dual requirement responds to the specificity and the complex "ecology" of graphene as a two-dimensional material, which appears in different configurations (graphite or diamond) depending on how the bonds are formed. The NGI was built "with other 2D materials in mind," because it is the interaction with other 2D materials that will determine the success of future devices (in membranes, in energy, in electronics).[21] The interactive capacity of graphene, and how it forms bonds, is what the institute is built around; similarly, the scientists and industry people in this building are expected to behave *like* graphene: isolation for the purposes of noninterference in experiments and interaction are the key conditions for the success of their work. Graphene sets the standards while at the same time researchers try to manipulate it in order to act according to internal and external constraints. And this process is performed incessantly through the searching for balance and delicate compromise. There is an amazing ontological symmetry that we witness: scientists, industry people, and 2D materials bear remarkable similarities to each other; they are all expected to form new bonds intensified by design. Thus, far from containing and sheltering scientific work, the building operates as a machine that maximizes the impact of graphene's implications and catalyzes the productivity of scientists and industry people, allowing for a faster exchange of ideas and creation of new ones. As graphene is evolving almost daily, there is a constant pressure for the building and its dwellers to adjust to ever-changing standards and the ecology of the graphene machine.

Two aspects that struck us in this observation are the two distinctive rhythms of the building related to two speeds: the "high-speed train"-like rhythm of development of graphene research related to a hectic pair of work and experimentation, and a demanding and highly competitive research culture (witnessed in the lab and seminars spaces); and at the same time, the contrasting, slow, calm, and easygoing rhythm of discussion and relaxation (witnessed in open, generously lit, often-empty atriums and breakout spaces). These two rhythms of dwelling in the home of graphene shaped the science lifestyles at the NGI. Let us now witness the two rhythms of the building that followed the operative logic of the graphene machine.

Rhythms of Dwelling in the NGI Building

At the NGI "the groups just grow and shrink very, very rapidly because it's very dynamic research."[22] This suggests an accelerated speed of activities. Everyone in the building follows the graphene rhythm—all the time, and in a hectic and devoted way. The building is self-contained, argues Vladimir Falko, research director of NGI, and its strength is that it has many complementary facilities for the fabrication of devices (electronics and opto-electronics, nanocomposites, and printable electronic systems) that are all invented within its space, characterized and then developed as prototypes for possible industry use. The swift pace of graphene research follows a tree-like development, the NGI being the start of this intertwined branching, extending to the Graphene Engineering Innovation Centre (GEIC) and other buildings. Facilitating collaboration, the NGI speeds up technology transfer through the coexistence of labs and industry partners, and also accelerates progress.

The building design matters for the rhythm of research. "Built with future development and expansion in mind," as John Whittaker, building manager of the NGI, tells us, its high-specification facilities are meant to provide "the flexibility for future development, as graphene 2D materials research will evolve over the coming decades."[23] Researchers constantly talk about potential, speed, and the possibility to adapt labs and upscale; they deal with projected changes, and think about long-term research developments and project management. Trying to anticipate what future research programs will bring, what the dwellers in the NGI witness is an "evolving building" whose flexibility is to be managed while its consistency is maintained.

One of the senior scientists, Mark Bissett explains that "scientists tend to be very territorial. We don't like sharing; it's a very competitive industry, especially in the case of graphene."[24] The competitive aspect of their work accelerates the rhythm of research and the labs are built to respond to this rhythm of development: they are flexible and could be used for different purposes; they allow easy connection to a gas line or to power through the gray spaces, and quick installation work. All this avoids wasting time and provides a smooth rhythm without disruptions to lab routines; as a result, the work tempo quickens and becomes more efficient. The in-house mechanical workshop avoids wasted time in subcontracting; the NGI technicians swiftly make changes on the spot. Even conventional design features, like the huge size of the elevator or the large width of the corridors are supposed to speed up the workflow. They try to respond to and predict future developments of technology and research pace. The elevator accommodates fifty-five people, and is designed with the forethought that heavy and bulky equipment might need to be moved swiftly from the storage space to the labs and installed. The corridors, unusually wide, are also designed to house such equipment; they will allow a certain number of palette tracks to be fitted and massive equipment to be wheeled straight into the labs or moved quickly into another lab if needed. The building is "very technology intensive," according to Fran Lopez-Royo, cleanroom manager, as a "high proportion of building volume is dedicated to the provision of technical facilities."[25] Anticipating future developments, the NGI facilitates the fast circulation of things and people throughout the building and remains at odds with the traditional buildings on the campus, not just in terms of aesthetic appearance and structure, but also, primarily, in terms of speed.

Another aspect related to speed is the extensive use of glass, which not only encourages communication and exchange of ideas, but also "makes it easier for incoming researchers to quickly get to know the experienced building users."[26] Transparency enhances knowledge of how to use equipment and dedicated spaces, and how to cope with the high-tech aspect of the building environment. Andrey Kretinin, lecturer in materials, and one of the permanent "residents" in the building, tells us that the open space design of the laboratories also matters for the speed of research. As he affirms, "I quite like it because the walls of my office are made of glass and I can actually see what my students are doing in the lab and you feel involved in the everyday life of your students."[27] The arrangement of offices in immediate adjacency to open space labs makes senior researchers more involved in the experimental work, which is very difficult in other, more traditional buildings on the campus. From his office, Andrey can access the lab immediately, take part in running experiments, share ideas, and adjust equipment. All this "actually speeds up the creativity process," explains Andrey. The "porous" infrastructure of offices and open labs affords "instant communication," contributing to an even more accelerated pace of research. On the one hand, this exposure between scientists increases the pressure while also optimizing performance. On the other hand, it makes working conditions more informal; it is easier to ask for help and engage in discussion. Expanding the overall "dwelling space" of scientists in the continuum open-lab-office also makes for a healthier, safer, and more spontaneous working environment in a compact building where all facilities are brought together and combined. As the building covers the whole process of scientific work, no time is wasted; there is no need to go out; everyone dwells in the enclosed, dense, and self-sustained territory of "graphene land."

Figure 3.7
Open labs and glass offices. Photograph by the authors.

The Breakout Rhythm

While there is no downtime for the accelerated speed of graphene's development and the NGI never sleeps, scientists do need downtime. As John jokes, the NGI is not "a complete slaughterhouse for research."[28] A number of spaces allow scientists to slow down the hectic rhythm of research and experimentation: the open terrace and atrium areas are quiet to allow some downtime, and a casual talk with friends. The corridors are lined in a black PVC material to encourage scientists to pause and write on the walls while chatting. To relax, Mark and Rahul often go upstairs to the roof terrace to sit down for a coffee, and to talk to people. Working in the building, Mark very often loses track of time: "the temperature control is very nicely done, so it's quite easy for me to spend a lot of time in the building. I forget how long I spend here, sometimes 8–9 working hours, and it never gets uncomfortable."[29] Other scientists also tell us that it is easy to forget what the weather is like when they do experiments in the labs that are isolated environments or contain commercially sensitive things and thus have no windows. According to Rahul, "in this building the social area is maybe more distinctive compared to the old buildings," and the attractive social aspects of the NGI encourage him spend a lot of time talking to colleagues.[30] For Andrey, the building affords physical and psychological comfort, reducing the enormous stress from teaching or research. Sarah and Rahul typically go into the building on weekday evenings for a couple of hours; Sarah even spends part of her weekends there enjoying the quiet and nicely designed atrium spaces. Here in the NGI everything is convenient, comfortable, close by, and accessible. The self-sufficient and comfortable environment, the pleasing modern aesthetics, the large open-air and green spaces, the abundant light, the atriums, the artworks, and the many common breakout spaces make the building a working paradise for all those residing in graphene land.

Thus, the building acts as a total environment where we witness an active process of dwelling at two speeds: hectic, hurried, and competitive, but also, slow, relaxed, quasi-domestic. Both the speedy and the relaxed rhythm encourage interaction and collaboration. No matter how intense the working rhythm is, scientists interact with each other, exchange tips, knowledge and lab recipes, rub shoulders at the benches and adjust experimental settings together in shared labs or in cleanrooms. Slowing down, they cross paths with each other and in this moment of unhurried, relaxed, breakout quality time, new ideas emerge, and new research alliances are formed that in return will accelerate anew the rhythm of the graphene machine.

Conclusions: The Good Experiment

The design of laboratory buildings has often been steered by narrow concerns for "efficiency." Praised for being a place where the instrumentalities of scientific research are shaped and where the credibility of scientific claims is secured, the contemporary laboratory is, however, far from a neutral "stage" for the production of knowledge.[31] Our study of the National Graphene Institute in Manchester shows convincingly that modern laboratory buildings are vital settings for the active shaping of new patterns of research cultures, new socio-technical ecologies of innovation, and new alliances of science, society, and industry.

Tracing the modalities of design dialogues between the architects and the scientific labs, and observing the rhythms of dwelling in the NGI, we showed that this high-tech nanoscience building provides an efficient—and most importantly—a pleasant and interactive working environment.

Figure 3.8a, b

(a, b) Atrium at the NGI. Photographs by the authors.

Figure 3.8c

(c) Nobel Laureate Sir Kostya Novoselov in the Atrium of National Graphene Institute. Photograph by the authors.

First, in our recollection of the process of making the NGI, neither the architect nor the lead scientist explained the building design by referring to their own individual idiosyncratic visions. Instead, different design options traveled many times between the architectural lab and the science lab, and were transformed and enriched as they traveled; both designers and scientists learned from each other's techniques. While the architect engaged in explanations about the nature of graphene, the lead scientist Kostya showed mastery of key architecture terms and concepts. Glimpsing the two labs' lives, we can conclude that, in an interesting and paradoxical way, the architectural lab appeared to bear more similarities with a scientific one than with the typology of the artistic studio, while the graphene lab stood out as reminiscent of a quasi-architectural setting where new types of research and partnership spaces were shaped. Graphene remained crucial for the design of the building, and its agency produced the drive for an ever-evolving building going always "beyond graphene," as Kostya repeatedly defined the ambition. Connecting the methods and techniques of the two labs, graphene's novelty spread through the networks of material science, chemistry, and engineering, and infiltrated back into the lab lives of both architects and graphene scholars. Thus, the NGI building followed logically, in the tangle of a tango, architectural constraints, and graphene standards.

Second, walking around the NGI and tracing the modalities of dwelling in the building, we can argue that the old-fashioned public image of scientists as recluses who work in isolation is far from the reality of contemporary science. Scientists interact constantly, collaborating and bouncing ideas back and forth. Science is extremely social, and the success of many innovations is directly proportional to the extent to which there can be good contact among different laboratories and different types of expertise. Yet, interaction is neither endless nor self-sustained, but rather artfully regulated by the NGI building. Affording two distinctive speeds—the hectic competitive rhythm of accelerated research and the slow, human rhythm of easing up, the NGI architecture acts as a flexible and porous setting mediating the balance between the regime of seclusion, silence, and isolation and the regime of interactive, noisy, vibrant social exchange. The success of scientific buildings today depends on how well they manage to regulate this precise balance. The NGI building acts as a guiding, daily reminder to practitioners not of who they are and where they stand, but of why they are there and what type of activity science is supposed to be. Its design does not shape identities; rather, it reshuffles practices, and redistributes agency. The NGI architecture underlines the visible steps in scientific endeavor, the formulae, the inscriptions and blackboards, the drafts and try-outs, the mistakes and retakes. All this tells us is: "There is no double click magic here! Come and try, make mistakes, start over: one formula after another on the blackboard, or on the building skin!" It is in this continuous dialogue between chalk sketches of formulae and engraved equations on a facade that design matters for graphene research, and enacts graphene's future.

Glimpsing the socio-material working dynamics in the two labs, and tracing the modalities of dwelling in the NGI, this chapter also outlines the importance of new theorizations of the simultaneously unfolding worlds of architectural and scientific experimentation and the related science lifestyles. The NGI building bears a striking resemblance to the settings of scientific experiments. Scholars like Andrew Pickering and Peter Galison have argued that the setting of a scientific experiment or a medical intervention can act as a mediator through which the object investigated gains capacities it did not have before.[32] The good experimental set-up, and by extrapolation, the good design, is not one that is passive, but rather one that deforms, constrains, and enables in thought-provoking ways. Isabelle Stengers defined a good experiment as an occasion for sudden reversals and unexpected results.[33] In a good experiment, the setting is not invisible and docile—it is present and active. In a similar way, we can argue that the NGI building sets an example of a good architectural experiment, and serves as an active agent in the transformation of scientific practices that are by definition interdisciplinary, and science-industry driven, and at the same time operates as evidence for these changes. Designed and experienced as an influential setting, the building "materially refigures" and actively transfigures the practices of all those who pass through it, no matter how different their ontology is—nanotubes and mega-machines, scientists and technicians, among others; it redistributes their agency in a slow and relationally efficient way. We witness the transformative effects of laboratory buildings as powerful material arrangements where objects, apparatuses, ideas, inscription devices, scientists, and materials such as graphene, are all submitted to tests and transformed as the building actively intervenes in their lives at different speeds. The modern laboratory building expands the boundary of a situated lab-based experiment, and acts as a powerful architecturally enhanced, mega-scale mediator, which does not simply contain, but also fabricates new capacities and multiplies the effects of science.

Notes

1. Bruno Latour and Steven Woolgar, *Laboratory Life: The Social Construction of Scientific Facts* (Beverly Hills, CA: Sage Publications, 1979); Karin Knorr-Cetina, *The Manufacture of Knowledge* (Oxford: Pergamon, 1981); Michael Lynch, *Scientific Practice and Ordinary Action: Ethnomethodology and Social Studies of Science* (Cambridge: Cambridge University Press, 1993).

2. Steven Shapin, "Placing the View from Nowhere: Historical and Sociological Problems in the Location of Science," *Transactions of the Institute British Geographers* 23, no. 1 (1998): 5–12; Sven

Dierig, Jens Lachmund, and Andrew Mendelsohn, eds., *Science and the City*, special issue of *Osiris* 18 (2003); Peter Galison, *Image and Logic: A Material Culture of Microphysics* (Chicago: University of Chicago Press, 1997).

3. Peter Galison and Emily Thompson, eds., *The Architecture of Science* (Cambridge, MA: MIT Press, 1999).

4. David Livingston, *Putting Science in Its Place: Geographies of Scientific Knowledge* (Chicago: Chicago University Press, 2003).

5. Thomas Gieryn, "City as Truth-Spot: Laboratories and Field-Sites in Urban Studies," *Social Studies of Science* 36 (2006): 5–38.

6. Michel Callon, "Le Travail de la Conception en Architecture," *Situations Les Cahiers de la Recherche Architecturale* 37 (1996): 25–35; Michel Callon, "Concevoir: Modèle Hiérarchique et Modèle Négocié," in *L'Élaboration des Projets Architecturaux et Urbains en Europe*, ed. Michel Bonnet (Paris: Plan Construction et Architecture, 1997), 169–174; Albena Yaneva, "Scaling Up and Down: Extraction Trials in Architectural Design," *Social Studies of Science* 35, no. 6 (2005): 867–894; Albena Yaneva, *The Making of a Building: A Pragmatist Approach to Architecture* (Oxford: Peter Lang Publishers, 2009); Bruno Latour and Albena Yaneva, "Give Me a Gun and I Will Make All Buildings Move: An ANT's View of Architecture," in *Explorations in Architecture: Teaching, Design, Research,* ed. Reto Geiser (Basel: Birkhäuser, 2008), 80–89; Sophie Houdart and Chihiro Minato, *Kuma Kengo: An Unconventional Monograph* (Paris: Editions Donner Lieu, 2009).

7. Dana Cuff, *Architecture: The Story of Practice* (Cambridge, MA: MIT Press, 1991).

8. Kenneth Frampton and Steven Moore, "Technology and Place," *Journal of Architectural Education* 54, no. 3 (2001): 121–122.

9. Reinhold Martin, "Architecture's Image Problem: Have We Ever Been Postmodern?," *Grey Room* 22 (Winter 2006): 6–29; Antoine Picon and Alessandra Ponte, *Architecture and the Sciences: Exchanging Metaphors* (New York: Princeton Architectural Press, 2003); Alejandro Zaera-Polo, "The Politics of the Envelope: A Political Critique of Materialism," *ArchiNed* 17 (2008): 76–105.

10. Gieryn, "City as Truth-Spot," 5–38; Galison and Thompson, eds., *The Architecture of Science*.

11. The university invested in a second dedicated building, designed by Rafael Vinoly, the GPB 60 million Graphene Engineering Innovation Centre (GEIC) due to open in 2018. To contextualize our study, we talked to a number of scientists and building managers related to the GEIC building and the graphene city agenda of Manchester.

12. Michael J. Bednar, *The New Atrium* (New York: McGraw-Hill, 1986); Albena Yaneva, "Is the Atrium More Important than the Lab? Designer Buildings for New Cultures of Creativity" in *Geographies of Science*, ed. Peter Meusburger, David N. Livingstone, and Heike Jöns (Dordrecht: Springer, 2010), 139–151.

13. Interview with Kostya Novoselov, Manchester, July 12, 2017.

14. Ibid.

15. Ibid.

16. David Taylor, *The Right Formula* (Manchester: Manchester University Press, 2016).

17. Interview with Kostya Novoselov, Manchester, July 12, 2017.

18. Interview with Tony Ling, Manchester, 2017.

19. Ibid.

20. Ibid.

21. Interview John Whittaker, Manchester, 2017.

22. Interview with Kostya Novoselov, Manchester, July 12, 2017.

23. Interview John Whittaker, Manchester, 2017.

24. Interview with Mark Bissett, Manchester, 2017.

25. Interview with Fran Lopez-Royo, Manchester, 2017.

26. Ibid.

27. Interview with Andrey Kretenin, Manchester, 2017.

28. Interview John Whittaker, Manchester, 2017.

29. Interview with Mark Bissett, Manchester, 2017.

30. Interview with Rahul Nair, Manchester, 2017.

31. Thomas Gieryn, "Two Faces on Science: Building Identities for Molecular Biology and Biotechnology," in *The Architecture of Science*, ed. Peter Galison and Emily Thompson (Cambridge, MA: MIT Press, 1999), 423–459.

32. Galison, *Image and Logic*; Andrew Pickering, *Science as Practice and Culture* (Chicago: Chicago University Press, 1992).

33. Isabelle Stengers, *The Invention of Modern Science* (Minneapolis: The University of Minnesota Press, 2000).

4   The Beach Boys: Classified Research with a Southern California Vibe

Stuart W. Leslie

Southern California's smartest defense intellectuals preferred to hang out in some of its toniest beach communities, at RAND in Santa Monica, Hughes Research Laboratories (HRL) in Malibu, and Nortronics in Palos Verdes—places designed to recruit, retain, and inspire them with bold architecture, scenic views, challenging problems, brilliant colleagues, and a lifestyle best described as "cold war avant-garde."[1] Long before companies such as Apple and Google learned how to attract and indulge their high-tech workforces with espresso bars, climbing walls, flextime, and other perks, these laboratories perfected the art of concierge science. Whether through curated contemporary art collections, guest lecture series by recent Nobel Prize winners, or simply the opportunity to live and play in some of the area's best beachfront resorts, these companies reimagined the scientific life as an aesthetic choice for members of an emerging "creative class," with a distinctly regional flair.[2]

In deliberate contrast to the button-down style of Ivy League universities, East Coast corporate laboratories, and Washington, DC think tanks, Southern California companies sold cool, casual, and quirky as the hallmarks of cutting-edge research. These "beach boys," as their often-envious counterparts called them, worked in strikingly modernist buildings with amenities more often associated in those days with country clubs than with research laboratories—tennis courts, swimming pools, putting greens, and al fresco dining. These laboratories appealed to scientific freethinkers who cultivated a serious interest in modern art, architecture, and music, and who had fashioned for themselves an image as visionaries freed from the constraints of conventional thinking, on and off the job. This strongly regional version of "suburban science" seemed worlds away from the middle-class conformity they associated with IBM, Bell Laboratories, and General Electric, where so many of them had begun their careers.[3] Southern California promised instead a new style of doing science, where researchers themselves called the shots, where the bottom line did not constrain blue-sky thinking, and where youthful exuberance had the chance to prove itself.

The scientists themselves would never have guessed how relatively brief their moment in the sun would be, made possible only by mushrooming federal defense contracts that could not sustain "fortress California" beyond the end of the Cold War.[4] Their parent companies would either close down these laboratories for good (Nortronics), sell them to new owners (HRL), or tear them down and rebuild them in ways that would be unrecognizable to their original employees (RAND). As the epicenter of high-tech glamour moved north, to Silicon Valley and companies that bet their futures on the consumer market rather than the military-industrial complex, Southern California's laboratory lifestyle seemed increasingly passé, a "blast from the past" that did not resonate with the expectations of a new breed of entrepreneurs or the young scientists who worked for and hoped to emulate them. The chic geeks did not appreciate how much their own style of science owed to an earlier version of California exceptionalism. For them, the beach boys seemed as outdated as the pop group itself: a 45-rpm record in an mp3 world.

Think Factory De Luxe

RAND's most famous scientist, Dr. Strangelove, never actually worked there, though he certainly could have been the fictional alter ego for several luminaries who did, including nuclear strategist Herman Kahn. In the 1964 film, Dr. Strangelove asked scientists from the "BLAND Corporation" to assess the Soviet doomsday machine, and contemporary audiences immediately caught the reference. RAND, like the doomsday machine itself, depended on people knowing just enough about it that it would be, in Dr Strangelove's words, "credible and convincing."[5]

Who could pass up the chance to work in a place with access to some of the nation's top secrets, influence at the highest levels of the military and government, and the opportunity to collaborate (and match wits) with some of brightest minds in select scientific fields? To attract the "permanent interest of scientific workers in problems of the Air Forces," and help stem the postwar exodus back to the universities, the Air Force came up with $10 million to fund a temporary home for RAND (an acronym for Research ANd Development) at Douglas Aircraft's Santa Monica headquarters.[6] For the right people RAND "would become a kind of secular monastery—worldly in rubbing shoulders with the physical and social sciences, industry, and the military, at all levels and all the way from Korea to London and the Continent; yet monastic in its security isolation."[7]

Over its first couple of years RAND quickly ramped up its staff to 200, first in the physical sciences, engineering, and mathematics, and then more gradually in economics and other social sciences. Strictly a blackboard and chalk operation without laboratories or machine shops, RAND issued several influential reports in its first years, including a famous study for the design of a satellite more than a decade before Sputnik.

In line with its maverick reputation, RAND tended to hire brilliant eccentrics like mathematician John Williams or physicist Herman Kahn, men too impatient to bother completing their own doctoral degrees at Princeton and Caltech respectively. Williams, longtime head of the mathematics department, became perhaps the quintessential RAND researcher. "Basically, we have the time and the inclination to think," he said. "Few people have both. We don't have to teach, design, build, operate, manage, legislate, or make decision[s]—except in-house. Our business is thinking, not doing."[8] The Air Force provided the startup money and some general guidelines about the problems it wanted RAND to think about, such as nuclear weapons design, strategic bombing, aircraft and ballistic missile design, and how best to fight a future nuclear war. To preserve RAND's intellectual independence, the Air Force kept it at arm's length from the defense establishment in Washington, DC. At thirty-seven-years old Williams was a gray eminence by RAND standards, while Kahn, just twenty-five, seemed no more precocious than many of his colleagues. Williams firmly believed that all work and no play would make RAND a dull place. He once sent a pointed "Note to a Workaholic," telling his colleague to "get the hell out and take some vacation. You are doubtless beginning to hear voices and your pallor is probably revolting to your neighbors."[9] Williams, who lived in the Pacific Palisades, spent his own downtime "souping-up" his Jaguar coupe with a Cadillac engine and roaring it up and down the Pacific Coast Highway late at night.

In 1948, RAND incorporated as a nonprofit corporation with a small endowment from the Ford Foundation, and moved from its makeshift offices at the Douglas plant to rented space in a Spanish-revival office building in downtown Santa Monica. As its staff swelled to 300 and began to spill over into surrounding buildings, RAND had to consider something more permanent. Williams, who had been with RAND from the very beginning, gave the matter serious attention and wrote up his ideas with a distinctive mix of wit, whimsy, and mathematical rigor. Recognizing that most RAND researchers sought—in order of priority—privacy, quiet, natural light, natural air, and spaciousness, Williams mathematically sketched out a design intended to promote RAND's interdisciplinary character. He noted that "it should be easy and painless to get from one point to another in the building; it should even promote chance meetings of people."[10] After taking account of a relatively compact site being considered opposite Santa Monica City Hall, Williams carefully calculated the required number and average sizes of offices, meeting rooms, and administrative spaces, and found himself drawn into "the theory of regular lattices, which is a fascinating subject; the square, the figure eight, and a hierarchy of more complicated designs."[11] He then worked out a table indicating the average travel distance between the points (offices) in different lattice configurations and arrived at a two-story arrangement of nine enclosed patios of specified dimension. "It might be that, in view of climatic conditions here, we could throw all or most of the halls out of the building," he wrote. "The patios could be surrounded by porches onto which the office doors would open. The porches would provide cover against the rain on those three or four happy days each year: otherwise, one could cross the patios from office to office. These small sheltered areas would not be windy."[12]

Williams made no pretense to being an architect, but RAND's architect H. Roy Kelley certainly appreciated the theory behind the lattice design. Kelley experimented with a number of "schemes," deploying open and enclosed patios of various numbers and dimensions before settling on a scaled-down version with four enclosed patios and one large open courtyard facing the beach. Designed as a modular system able to grow with RAND, Kelley's initial plan was completed in 1953 and then expanded by him in two subsequent additions that filled in an eight-patio matrix. By design, the checkerboard or "waffle" layout of the building encouraged serendipitous encounters of researchers from different disciplines who might find themselves sharing lunch or coffee in one of the patios.

To fill those buildings, RAND recruited people who would respect its fiercely egalitarian culture established by intellectual gunslingers who measured their worth by merit and moxie rather than job title. Department heads were told to search out people smarter than themselves on the theory that even a genius surrounded by yes men would never match a less gifted scientist with truly talented staff reporting to him. Initially recruitment would be tough, since RAND had no track record, no tenure, and few academic perks. They offered salaries slightly higher than the best universities and slightly lower than the defense industry, calibrated, in the spirit of a place that had perfected game theory, so that half of the candidates would ultimately accept. RAND generally made offers to the top 10 percent of graduating PhDs in any one field, no more than six or seven in any one department a year. It then brought the best of them out to Santa Monica for interviews, usually in the winter months. Its early advertisements prominently featured photographs of Santa Monica's beach. Still, RAND's real draw was its acknowledged leadership in emerging fields such as game theory, operations research, and rational choice theory, alongside the chance to work not only with RAND's stellar staff, but with its even more stellar consultants such as John von Neumann, Kenneth Arrow, and Paul Samuelson, among others.[13]

Figure 4.1
RAND Corporation's waffle building in Santa Monica, California, in the early 1960s. Photograph courtesy of the RAND Corporation.

East Coast transplants such as Albert Wohlstetter, who earned his graduate degree in mathematical logic at Columbia University and joined RAND in 1951, enthusiastically embraced the West Coast lifestyle. Wohlstetter was best known for such studies as "Selection and Use of Strategic Air Bases" (with its idea of "fail-safe") and his highbrow taste.[14] For his personal residence, Wohlstetter selected a secluded lot in the fashionable Laurel Canyon. Architect and next-door neighbor Josef van der Kar, who chose only the most discriminating clients, designed the Wohlstetter House in a California Modernist style. Noted landscape architect Garrett Eckbo, another neighbor, designed the pool, with boomerang benches and a Mondrian-style screen, as well as a zigzag lawn and a desert garden.[15] Wohlstetter often threw lavish office parties at the residence, featuring live chamber music that at times was performed by RAND employees. No wonder a *Life Magazine* photographer sent to capture the mystique of RAND's "valuable

Figure 4.2
RAND scientist Albert Wohlstetter's California Modernist home in Laurel Canyon, by architect
Josef van der Kar. Photograph by Julius Shulman, courtesy of the Getty Research Institute.

batch of brains" shot an after-hours bull session in Wohlstetter's den.[16] On salaries of
$25,000 or less, most RAND PhDs had to settle for Santa Monica or West LA, though
many aspired to the high style of the Hollywood Hills.

Back at work, RAND scientists took full advantage of its seaside setting and patio
layout. Whether they preferred ping pong, shuffleboard, putting practice, *Kriegspiel* (for
"war game," the blind version of chess favored by the German high command and by
young mathematicians from Princeton), or merely al fresco lunch under the umbrellas, the patios became the building's social hubs. The pink and white headquarters
reminded a visiting reporter of "a university without students—a university in which
the whole faculty decided to come in on its day off. Men in garish sport shirts stroll
casually about; nobody seems to hurry. Uniforms are rare; the few Air Force officers on
temporary duty show up, like the permanent staff, garbed in casual Southern California

comfort."[17] Lunchtime swims or surfboarding were nothing out of the ordinary either. Perhaps only RAND would have allowed one of its researchers to spend time on a scientific study of surfing and then let him patent the idea.[18]

For a place that prided itself on unconventional thinking, RAND seemed an ideal sponsor for LACMA's Art and Technology Program, a pursuit that placed avant-garde artists in residence with high-tech companies in the Los Angeles area. Conceived in 1966 by Maurice Tuchman, LACMA's first curator of contemporary art who had recently arrived from the Guggenheim, the idea was to encourage young artists to work in new media and in stimulating new environments, and of course to encourage corporations to think of themselves as patrons of contemporary art—RAND jumped at the opportunity.[19] Artist in residence John "Crashed Cars" Chamberlain, a sculptor, figured out how to engage, if often enrage, RAND's staff, deliberately assuming the role of intellectual provocateur so much in the RAND tradition. He screened one of his explicit experimental films in the cafeteria. He chain-smoked and held court at one of the patio tables outside his office. He peppered the staff with memos asking them for "ANSWERS. Not Questions!," which he then compiled into "Rand Piece," a perplexing bit of guerilla art. Baffled, RAND researchers responded in kind: "Quit Wasting RAND Paper and Time. The Air Force needs thinkers—where do you fit?"[20] Paradoxically enough, RAND's raison d'être was thinking up answers that spoiled conventional questions, but not answers to the kind of questions Chamberlain had in mind.

Software entrepreneur Peter "Utilities" Norton introduced RAND to contemporary art in the more traditional way; by hanging a significant part of his growing collection on its walls. Norton and his private curator selected sixty pieces for RAND and installed them on what had been largely barren walls in the offices, lobbies, conference rooms, and corridors. The 1991 catalog for this collection, heavy on Derrida and deconstruction, must have seemed as opaque as anything Chamberlain could have come up with.[21] The collection itself was a hit, featuring several pieces by Andy Warhol and three others by local favorite David Hockney, including a photo-collage taken in Santa Monica.[22]

By the mid-1960s, some RAND veterans wondered if the place had lost its edge. Some of the mavericks had moved on; Herman Kahn to the new Hudson Institute, and Albert Wohlstetter to the political science department at the University of Chicago. In the Vietnam War era, RAND more often seemed to be the problem rather than the solution, its strategic doctrines increasingly out of step with the political realities of a new phase of the Cold War. Daniel Ellsberg personified RAND's dilemma. Trained at Harvard in economics and decision theory, Ellsberg became a RAND wunderkind in the mold of Herman Kahn, and later an influential adviser to the Pentagon and the State Department. In 1967 Ellsberg returned to RAND with the assignment of compiling a classified history of the Vietnam War. Disillusioned by what he learned during the study and by the war itself, Ellsberg secretly copied and released the classified documents entrusted to him. The *New York Times* subsequently published them, with commentary, as the politically explosive "Pentagon Papers."[23]

RAND meanwhile deliberately diversified its research portfolio from classified military projects to studies better suited to the funding priorities of the Great Society—urban affairs, transportation, housing, and healthcare.[24] Even admirers agreed: "As the 1970s wore on, RAND shifted more and more from actively creating the frontiers of technology to concentrating on policy analysis, military and nonmilitary."[25] Much of its early aura as a place of dark secrets vanished. To critics on the left and on the right, RAND seemed headed toward becoming "just another outfit running after contracts," in a building as dated as its reputation.[26]

Shangri-La for Deep Thinkers

Hughes Research Laboratories in Malibu could be as secretive, glamorous, and seductive as the reclusive billionaire Howard Hughes himself. Could there be any more appropriate setting for his namesake laboratory than a glistening aerie overlooking the Malibu movie colony? No wonder comic master Stan Lee modeled the original *Iron Man* (1963) on Howard Hughes. The film version set Iron Man's clandestine laboratory in the basement of a Malibu mansion freely borrowed from some of architect John Lautner's futuristic follies.[27]

As an aviator and engineer, Hughes may be best remembered for his sensational if failed prototypes: the H-4 Hercules ("Spruce Goose"), the largest transport plane ever built, and the XF-11, a photo-reconnaissance plane he personally crash-landed into a house in Beverly Hills on its first test flight. During the war, Hughes purchased a large tract of land in Culver City, just north of the Los Angeles airport, and constructed an industrial complex where Hughes Aircraft built planes under contract to other airframe manufacturers. With characteristic business savvy, he decided that rather than go head-to-head with established Southern California firms like Douglas, Lockheed, or Northrop, Hughes Aircraft would look for a niche it could profitably exploit. As the brain trust for a serious R&D program, Hughes hired two brilliant electronic engineers, Simon Ramo and Dean Wooldridge, Caltech classmates who had gone on to successful careers at General Electric and Bell Labs respectively. Though small, the Hughes electronics division had one decisive advantage, the essentially unlimited capital of its sole shareholder Howard Hughes, who could afford to take the long view. Ramo and Wooldridge also had the confidence of the young Air Force officers in charge of its avionics and missile programs, who shared their prejudice that traditional airframe manufacturers could not attract and hold truly top-notch scientists and engineers.[28]

The duo set out to build the Bell Labs of the West, offering top salaries, compelling technical challenges, and, as the recruiting posters promised, "luxury living in California," poolside beneath the palms.[29] What Hughes Aircraft could not offer, given its single private owner, was any equity in the company. Nor, despite its advertisements, did it provide any architectural inspiration. The Culver City complex had a grim factory aesthetic, with every building painted a "sea mist green" specified by Howard Hughes himself. In the summer of 1953 Hughes's eccentric and largely absent management style boiled over into a wholesale defection of top scientists, engineers, and executives, led by Ramo and Wooldridge themselves. Despite the losses, Hughes Aircraft rapidly recovered once its founder stepped aside at the insistence of the Air Force and "donated" the company to the Howard Hughes Medical Institute as a tax shelter. Under new president Leland Hyland, a former Bendix executive, Hughes Aircraft became the largest military electronics firm in Southern California, and one of the most innovative.[30]

Hughes Aircraft, with no stockholders to answer to, offered its scientists and engineers remarkable latitude to pursue blue-sky ideas. For someone like Harold Rosen, who designed the first generation of geosynchronous communications satellites, the company offered an ideal environment. "About the time I was making my choice [between graduate school at Harvard and Caltech]," he remembered, "the old *Life* magazine … came out with a big story on beach parties in southern California, and I decided I'd prefer that than the cold New England winter! So I decided to go to Caltech."[31] He joined Hughes Aircraft in Culver City where he went to work on radar-guided missiles. Rosen and his two principal colleagues, all enthusiastic surfers and scuba divers, recalled getting a key insight for the satellite's positioning jets from watching jellyfish swim.[32]

In 1959, following the cancellation of a major Air Force contract, and contrary to conventional wisdom, Hughes Aircraft tripled its R&D budget. It also decided that its best researchers required some distance, intellectual and geographic, from the day-to-day pressures of advanced development projects if they were going to come up with the next big breakthrough. To house them, the company identified a property above the beach in Malibu, built but never occupied, by the Potter Aeronautical Corporation, a small New Jersey firm that manufactured flow meters for rocket engines and had expansive plans but limited capital.

Architect Ragnar Qvale had designed for Potter a Y-shaped building with unrivaled views of the Pacific Ocean to one side and the Santa Monica Mountains to the other. Qvale was a respected residential architect with matinee idol looks and a resume to

Figure 4.3
Aerial view of Hughes Research Laboratories above Malibu, California, 1959, designed by architect Ragnar Qvale, with full glass walls for beach or mountain views. Courtesy of HRL.

match, including the original Sahara Hotel in Las Vegas and the local Wilshire Country Club.[33] His flat, two-story structure for Potter Aeronautical hugged the Malibu hillside, with twin wings curving off the central spine like the Greek letter 'Ψ' (psi). Entirely sheathed in glass, with a cantilevered canopy for shade in the intense Malibu sun, the building featured an enormous fieldstone wall near the entrance, and a bridge of inter-locking concrete slabs crossing a large landscaped pond. The fieldstone extended to the left side of the entrance and into the lobby, a contrast against the dominant white concrete and glass motif.

Only the exterior had been completed, so Hughes Aircraft fitted out the interior itself, with offices along the exterior walls—each with a million-dollar view—and labo-ratories along the interior corridors. The scientists had to pay for those views, however; partitioned on the seven-foot wide window frames, the offices could be monastically claustrophobic. Cozy could also mean collaborative, though, as researchers had every reason to get out and take a walk around.[34] Qvale had never completed the planned swimming pool and cafeteria, but the sixty-acre grounds offered plenty of paths for contemplative strolls and shaded benches for lunchtime conversation. Being in one building "brought us physically closer together and made casual interactions easier than they were back in Culver City," one researcher reminisced.[35] Such an isolated facility had to be as self-sufficient as possible, with its own draftsmen, machinists, and glassblowers to custom-tailor experimental apparatus. The shop, with its long row of polished machine tools, got the same inlaid pecan floors as the rest of the building, accompanied by the same spectacular views. The lobby, furnished in Danish modern, looked like the entrance to some exclusive country club or casino, precisely what Qvale had in mind.

Hughes Aircraft moved its research laboratories into their new home in 1960. Malibu had limited and pricey housing; perhaps a third of the staff chose to live there anyway for its resort-like flavor. Most of the other employees lived down the coast in Santa Monica or the Pacific Palisades, though some drove in over the mountains from the San Fernando Valley, as one individual noted: "Sure it's a long haul. But it's such a darn pleasant place to work I actually look forward to the drive every morning."[36] Like RAND, Hughes sold sun, sand, and surf to winter-weary graduate students from the best programs in the Northeast and Midwest. Thanks to popular music and films, Malibu became synonymous with Southern California youth culture, and Hughes's researchers could often be found after hours at one of Malibu's beachfront bars, or on their boards.[37]

Hughes Research Laboratories (HRL) repaid the company's investment almost immediately. Theodore Maiman fired up the world's first functioning laser there in May 1961, beating out Bell Labs and its high-powered team of future Nobel laureates, bringing the new laboratory international recognition.[38] HRL sought to retain a university atmosphere even in a highly classified facility. They recruited heavily from Caltech, especially at the PhD level, and many of those alumni moved to the upper ranks of HRL. To keep some of that freewheeling graduate school spirit alive, HRL invited Richard Feynman, the theoretical physicist, raconteur, and keen popularizer of science, to give a series of weekly lectures for the staff. For five consecutive years Feynman gave Monday afternoon seminars, with open invitation. Most attendees suffered from what they dubbed the Feynman Effect: "Around halfway through the lecture he would reach a depth I couldn't follow anymore and I would have to bob up to the surface for air while he continued descending into the murky depths of science and math."[39] What Feynman said probably contributed very little to HRL's technical accomplishments, but where and how he said it added immeasurably to the esprit de corps.

HRL advertised for researchers fascinated by "far out" ideas, and attracted more than its share of "visioneers."[40] After Arthur Chester became HRL director he thought some contemporary art might add a touch of inspiration to the halls of the laboratory. When Atlantic Richfield (ARCO) closed its Los Angeles headquarters and auctioned off its contemporary art collection, Chester bought some of the larger pieces and hung them on HRL's walls. For the lobby, he commissioned a signature sculpture from Eugene Sturman, a Malibu-based artist whose monumental *Homage to Cabrillo* is downtown LA's official time capsule. HRL researchers then had the chance to evaluate Sturman's maquettes and select their favorite, which Sturman subsequently scaled up into a finished piece.[41]

HRL became the public face for a company with a reputation for keeping a low profile. For generals, visiting executives from other companies, and scientific dignitaries alike, it was one place everybody wanted to see for themselves. In 1965 Hughes Aircraft invited the European press corps to cover the upcoming launch of Intelsat 1, the first geosynchronous communications satellite. Hughes Aircraft management made sure the journalists saw two Southern California landmarks on their trip: Disneyland, with a personalized tour by multilingual guides, and HRL, whose own imagineers talked about lasers, ion propulsion, and other real-life science fiction.[42]

The Perfect Place to Think

Where other Southern California laboratories brought the scientists to them, the Nortronics division of Northrop brought its laboratory to the scientists. It established its research center in the heart of Palos Verdes, adjoining some of the most desirable residential real estate in southern Los Angeles, where many of its best researchers already lived.

Jack Northrop learned aeronautical engineering at the school of hard knocks, beginning as a draftsman for Lockheed in 1919 and working his way up to chief engineer by 1929, with a few years in between as a project engineer for rival Douglas Aircraft. He could count among his designs such classics as the Lockheed Vega, Amelia Earhart's record-setting airplane. Northrop, a restless visionary in the Howard Hughes mold, never fit comfortably into a corporate hierarchy and founded (and subsequently folded) several companies in pursuit of his unorthodox designs. In 1939, he incorporated the Northrop Corporation in Hawthorne, southeast of the Los Angeles airport. Northrop ran his company more like a "Skunk Works" than a conventional airframe

Figure 4.4
Two HRL scientists stroll along the dramatic exterior of the Malibu laboratory. Courtesy of HRL.

manufacturer.[43] His obsession was the flying wing—an airplane without a conventional fuselage and tail—among the boldest of aircraft ever envisioned. It was conceived, however, decades ahead of the control technologies that would have allowed them to be piloted safely. Northrop, embittered by the cancellation of his pet project, abruptly resigned from his own company in 1952.[44] Northrop without Jack, like Hughes without Howard, grew into a financially successful defense contractor, though known for solid engineering rather than revolutionary innovation. By the late 1950s Northrop's mile-long manufacturing facility in Hawthorne employed nine thousand "Norcrafters," relatively few of them scientists or engineers.

If Hughes Aircraft was a laboratory looking for a factory, Northrop was a factory looking for a laboratory, one that could help it win future systems contracts essential for high-performance weapons. To compete with Hughes, TRW, and other systems firms, Northrop established a Nortronics division at its Hawthorne facility in 1957. Its name and its logo—the characteristic sine wave of a radar screen—signaled a new corporate direction. Nortronics advertisements asked graduating engineers at places like Michigan Tech if they would like to work in the "electronic, aircraft/missile center of the world. … where you'll be able to spend your leisure at the Pacific Beaches, in the mountains, on the desert. You'll enjoy an active life in Southern California's incomparable year-round climate."[45] Add in a top salary and the chance to earn an advanced degree at company expense, and such an offer must have been all but irresistible to students stranded on the frozen tundra of Michigan's Upper Peninsula. Industrial Hawthorne, however, hardly evoked the image of Southern California Nortronics had in mind, though the Beach Boys did tape their first hit, "Surfin," in the Wilson family's Hawthorne ranch house.

To find a place where "our men can think and work away from the annoying distractions of a noisy industrial complex," Nortronics scoured the California coast, from San Francisco to San Diego, and settled on Palos Verdes, an upscale community just ten miles from its main factory.[46] Nortronics's timing could not have been better. Architect and mega-developer Victor Gruen, who literally wrote the book on *The Planning of Shopping Centers*, had just completed the master plan for a 6,000-acre project on the Palos Verdes Peninsula.[47] In Palos Verdes, Gruen had an opportunity to plan an entire town, with a shopping center at the foot of the enormous property, and offices, apartment buildings, single-family homes, medical center, and a golf course climbing to a resort hotel at the crest.

What distinguished this Gruen master plan from all the others was that it included a 400-acre research park zoned strictly for research and development. Great Lakes Properties, which owned the land, teamed up with General Telephone's industrial development department for a national marketing campaign aimed at "making science feel at home in California."[48] As planned by Gruen, the research park would include one ninety-acre site for a single large tenant, a dozen two- to ten-acre sites for smaller companies, and finally "research group facilities," a proto-incubator for startups, clustered around meeting rooms, restaurants, tennis courts, parking lots, and a shared auditorium. To protect property values throughout the development, the research park had similar restrictive covenants on density, architecture, and landscaping.[49] In one advertisement, a sign to the Palos Verdes Research Park pointed toward the sun-drenched (and smog free) cliffs above the blue Pacific.

Nortronics hardly needed convincing. The research park was only a short drive from its Hawthorne factory, and an even shorter drive for all the Nortronics scientists and engineers who already lived in Palos Verdes, including its general manager and several other high-ranking executives. The area offered an attractive alternative to the more expensive aerospace suburbs to the north, and its residents included growing numbers of physicists, engineers, and mathematicians commuting to jobs at Hughes Aircraft, North American Aviation, Douglas, and Northrop.[50] Paul Revere William, "architect to the stars," lowered his sights, or at least his prices, for SeaView, a 190-unit tract in Palos Verdes. Its models—the Lido, the Monte Carlo, the Bermuda, the Eden Roc—conjured up resort living on a VA-loan budget. Veterans purchased 95 percent of SeaView's low-slung ranch houses, each featuring "delightful luxuries which make the difference between 'a place to live' and a residence you enjoy to the full and want others to see."[51] Starter homes in developments bordering the research park—"where prestige wears a new low-price tag"—offered models (the Capri, Dauphine, and Shelter Cove) aimed at young up-and-comers.

Nortronics had its site. Now it needed an architect worthy of it, and found this in Charles Luckman. With one-time partner William Pereira, Luckman had designed some of the most stunning aero-spaces in Southern California. For less than $3 million, Luckman proposed a six-building campus (including a small observatory) for Nortronics, to be completed within two years for the bargain-basement price of $16.50 per square foot.[52]

Seeing no point in trying "to compete with the natural beauty, the mountains, the ocean and the expanse of the site," Luckman came up with a suite of four single-story buildings balanced along a central axis. To give them a little zip he topped them with eye-catching folded-plate concrete shells, run down the centers of their otherwise flat, steel rooftops.[53] Besides adding some visual panache, the folded-plate roofs opened space for a utility core above the laboratories, making them extremely flexible. Luckman set the buildings on slightly different elevations to add to the campus feel. To shade the buildings and to give them a more spacious quality, Luckman added six-foot-wide concrete overhangs along the perimeters.

Nortronics executives insisted on visibility and "a commanding view of the ocean," to which Luckman responded by designing a two-story administrative building at the far end of the campus, with a conventional flat roof and metal lattice "eyelids" shielding its glass curtain wall. Encircled by twin reflecting pools that doubled as an emergency water reservoir and reached by an elevated causeway, the administrative building had undeniable elegance. The entire complex was intended as an architectural expression of the Nortronics slogan, "Geared to the Space Age."[54] Tellingly, Luckman's architectural rendering included a photograph of a spectacular home overlooking the Pacific Ocean, accompanied by a whimsical map of Palos Verdes with Nortronics at its center, surrounded by small cartoons of people fishing, hiking, golfing, sailing, scuba diving, water skiing, and horseback riding—living the California dream.

Figure 4.5
Architect Charles Luckman's campus for Nortronics on the Palos Verdes Peninsula (1961). Photograph courtesy of California State University Dominguez Hills, Archives and Special Collections.

If Luckman could not compete with the natural splendor of the site, he certainly thought he could enhance it. He hired a prominent Beverly Hills landscaping firm to sculpt rock gardens, pools, and tree-covered knolls connected by footpaths. He put the parking lots out of sight by terracing them into the hillsides. As a later general manager freely acknowledged: "Sure this place is extremely impractical. Just look at the layout. It certainly isn't very good for P and L (profit and loss), but it brings in the talent. We have no trouble trying to hire here."[55] And the people Nortronics hired more often than not moved to Palos Verdes; more than half of its total workforce lived there.

Nortronics and its six hundred employees, including four hundred scientists and engineers, moved into their new headquarters in April 1961, and specialized in highly classified inertial navigation and guidance systems for bombers, submarines, and missiles. These systems demanded extreme levels of mechanical and optical precision, measured in millionths of an inch or in seconds of arc.[56] In Nortronics cleanrooms—oddly claimed to be "far cleaner than any housewife's kitchen"—highly skilled machinists and opticians turned blackboard calculations into working hardware. Nortronics did in fact hire a housewife from nearby Torrance for one of its unique projects, however; her voice became an automated warning system for bomber and submarine crews, who seemed to respond better to a "calm female voice" than to any light, bell, or buzzer.[57]

Employees called Nortronics "The Park," and from the road the place certainly gave the impression of a private golf course. The otherwise Spartan offices, laboratories, and shops came with country-club views. At lunch time the "serious-looking men staring at diagrams or writing on blackboards" could be found playing pitch and putt-golf on the well-trimmed greens, or, for a younger generation, throwing Frisbees.[58] Having a golf cart for the corporate mail run seemed only fitting.

For Nortronics the biggest challenge was living up to the covenants on the property, which blocked it from fabricating anything more than prototypes and set arbitrary

Figure 4.6
The administration building for the Nortronics campus offered ocean views and a golf club ambience, with its own putting green. Photograph by Julius Shulman, courtesy of the Getty Research Institute.

weight limits on even those. When Northrop, seeing important advantages in co-locating its R&D with high-tech manufacturing, and with plenty of land to spare on its original parcel, asked for a variance so it could construct additional buildings for small batch assembly, it immediately ran into fierce opposition from surrounding community associations. This defiance stemmed from a fear of even the lightest industry bringing low-cost housing in its wake. The same restrictive zoning made other potential tenants think twice about relocating in the research park, and after five years Nortronics found itself the one and only tenant. With land values on Palos Verdes rising so rapidly, real estate seemed a better investment than R&D, and local municipalities annexed the research park, excluding the Nortronics parcel, for residential development. Nortronics, with no reason to hold on to such valuable land, sold all but thirty-two acres to homebuilders. Nortronics remained "the perfect place to think," but found itself increasingly isolated, even alienated, from the very community where 260 of its own employees lived.

Conclusion

RAND got an unexpected opportunity to reinvent and reinvigorate itself after the Northridge earthquake of 1994, which badly damaged its Santa Monica headquarters. Seriously overcrowded for decades, RAND had considered relocating or rebuilding several times. After exploring a number of options—Washington, DC (near its major clients), North Carolina's Research Triangle Park, Cambridge, Massachusetts, Palo Alto, and even Pittsburgh—RAND decided that uprooting would be too costly, both in financial and cultural capital. With increasingly pricey Santa Monica real estate as its major financial asset, RAND agreed to sell all but a 3.4-acre parcel at the southern end of its fifteen-acre site to the city for $53 million, about half the anticipated price tag for its building.[59]

RAND expected its new headquarters, like its predecessor, to be a billboard for "an organization known for problem-solving," though with a very different personality. Several Pritzker Prize-winning architects competed for the commission, including I. M. Pei, Richard Meier, and favorite son Frank Gehry, whose Binoculars Building in Venice was a local landmark. RAND opted instead for Los Angeles-based DMJM, a more conservative firm with plenty of experience designing functional corporate headquarters. Seeking to recapture some of the old magic of the original waffle layout and to get some buy-in from an often-skeptical staff, RAND's management undertook a lengthy series of surveys, seminars, assemblies, and one-on-one meetings with almost every employee, all aimed at soliciting a "wish list" for the building. In addition to

Figure 4.7
RAND Corporation's new headquarters building in Santa Monica, just south of the original waffle building. Designed by DMJM and completed in 2005. Photograph by Benny Chan (Fotoworks). Courtesy RAND Corporation.

predictable recommendations such as "collegial," "flexible," and "interactive," the researchers also asked for "funky" and "a Buick without leather seats"; surely some indication of an aging staff![60] Inspired by one of John Williams's original matrix models, the figure eight, DMJM came up with two five-story opposing and intersecting arcs, like the outline of a fish, with sky bridges and courtyards linking the twin arcs at every level.[61]

Nothing could mask the new building's clear corporate character, anything but "funky" in its fit and finish. By administrative fiat, employees had to choose from preselected suites of office furniture—no more comfortable couches or easy chairs from home. The staff did get meeting rooms, a cafeteria, and a theater-style auditorium worthy of a Fortune 500 company, though, and the lushly planted and gorgeously lit courtyard could handle the most elegant social function day or night. No putting greens, ping-pong tables, or shuffleboard courts here; and no clutter, though the upgraded yellow umbrellas recall days past. These walls give the Norton art collection the museum-like quality look it deserves.

Would a genuine eccentric like John Williams or Herman Kahn ever feel at home in RAND's new headquarters, dominated as it is by earnest young policy analysts indistinguishable from their DC counterparts in similar think tanks? RAND may now be a more comfortable place to work, but not necessarily one that will encourage the intellectual audacity that once made it famous. In the corner of the sleek wood-paneled lobby, under glass, is a tribute to the golden age: a surprisingly faithful paper model of the original waffle building lovingly constructed by longtime RAND economist Tony Pascal. Playfully placed are some "wise old owls" in the patios, and like a medieval reliquary in the form of a cathedral, Pascal's model provides an enduring reminder of why someone went to all the trouble in the first place.

HRL outlived the company that created it and has managed to thrive despite repeated downturns in Southern California's aerospace industry. In 1984, following a long legal battle, a federal court ordered the Howard Hughes Medical Institutions (HHMI) to divest its holdings in the Hughes Aircraft Company and reinvest the income in a more diversified portfolio. HRL became a jointly owned venture of General Motors, Raytheon, and, later, Boeing.[62] HRL's relative isolation, small size, and divided ownership enabled it to remain remarkably independent of its corporate bosses, whose light touch allowed it to operate like a boutique laboratory.

HRL never entirely forgot the original Hughes Aircraft strategy of building a factory for its scientists and engineers; they designed gallium-arsenide circuits reliable and rugged enough for Hughes satellites and the unforgiving environment of space, and about the only way to manufacture exotic semiconductors to precise military specifications was to do it yourself. So, in 1987, HRL hired A. C. Martin Partners to design a $34 million addition, Building 254, where HRL could fabricate the $10,000 per piece circuits demanded for a $100 million spacecraft. The five-story addition, connected to the bottom corner of the Y-building's stem, could not match the elegance of the original. From the wrong angle its cascading decks and horizontal railings made it look like a cruise ship marooned on the hillside. Building 254 finally gave HRL some of the amenities a new generation of researchers had come to expect, however: a proper cafeteria instead of a food truck, an espresso bar, a fitness facility, and even a Wii room for the video gamers.

HRL may have lost some of the luster of its golden age, but it still attracts some top talent. In ultra-high-performance circuits, micromaterials (including the world's lightest lattice), computer science, and image recognition, HRL more than holds its own. There are fewer skilled machinists nowadays, and they do not wear bow ties, have ocean views, or a pecan parquet floor, but the current crop of scientists and engineers choose HRL for many of the same reasons as their predecessors: world-class science, and a great location. Almost everyone mentions the ocean views and gorgeous sunsets, though the occasional killjoy rips HRL as "an old aerospace company run like it was still 1965."[63]

Meanwhile, the old Culver City campus has gotten an extreme makeover and become home for a new generation of high-tech firms. The Ratkovich Company bought the property in 2010 and invested $50 million to renovate its eleven landmark buildings (including the gigantic wooden hangar where Hughes built the Hercules H-4), and to provide the kind of high-style environment its original scientists would never have envisioned.[64] With the satellites of Facebook, Yahoo, Google, and Microsoft as tenants, along with outposts of local academic powerhouses—USC's Institute for Creative Technologies and UCLA's IDEAS incubator—this high-tech ensemble named The Campus at Playa Vista has already been dubbed "Silicon Valley South."[65] The Campus, in turn, has spurred commercial, retail, and housing development in surrounding neighborhoods, driving up rents and driving out longtime residents of once working-class communities.[66] That the giants of new media have displaced what had been the largest aerospace employer in Southern California seems only fitting now that the entertainment industry has surpassed aerospace as the region's top industry in total jobs and export value. The hangars Howard Hughes once haunted now house Hollywood sound stages. Martin Scorsese passed up the chance to film his biopic *The Aviator* there, but Marvel Studios shot *Iron Man* at Playa Vista. What better nod to the man who started it all?

Nortronics, landlocked by the residential communities rising around it, could see no real future in Palos Verdes, and sought to cash in on the real estate bonanza it helped spawn. The land alone had quickly become worth far more than what Nortronics paid for it, even after adding the cost of the $4.4 million campus. Of course, only very select buyers would be interested in such a property. By far the best prospect was the newest branch of the California State University (CSU) system, South Bay State College. Palos Verdes seemed ideal, with a large, well-to-do population that had been complaining for years about a lack of local higher educational opportunities. The Nortronics site offered breathtaking vistas, sufficient land to build out a new campus, and ready-made buildings for the science departments.[67]

Anticipating a fast track, CSU hired architects A. Quincy Jones and Frederick Emmons to draw up a master plan and submit a design for the campus itself. They came back with a conventional mid-century modernist block of low, flat-topped buildings with open-air atriums. What killed the idea was opposition from community associations, no happier with the prospect of a university in their midst than a high-tech manufacturing complex. Rather than pick a fight with the neighbors, CSU held classes in temporary buildings. In 1965, Governor Pat Brown, looking to rebuild a community hit hard by the Watts riots, chose a permanent site for CSU Dominguez Hills in Carson, where Quincy and Emmons designed an appropriately brutalist campus.

Nortronics tried and failed to sell the property to redevelopers several times, including a near miss with Ratkovich, the eventual developer of the Hughes Culver City complex. Ratkovich thought that a remodeled and expanded Luckman campus could be transformed into a showpiece that would "bring people closer to their place of work," especially "residents of the peninsula now working for South Bay companies in aerospace, electronics, engineering and other high technology firms"— precisely what the research park had promised in the first place.[68] In 1991, having decided, ironically enough, that "skyrocketing home values have made it difficult to recruit engineers," Nortronics moved the last of its 100 scientists and engineers back to Hawthorne and put the property on the block.[69] An Irvine developer finally bought it in 1996 for $12.5 million, tore down the old laboratories, and replaced them with Vantage Pointe, a gated community of sixty-eight homes generously described as "traditional rural and country French."[70] Young aerospace engineers, with none of the job security of their predecessors, may have aspired to the same kind of lifestyle but could no longer afford it.

Neither could the companies that employed them. Faced with drastic cuts in defense spending in the early 1990s, Southern California aerospace companies had to pay closer attention to the bottom line than in their Cold War heyday of cost-plus contracts. The long-standing assumption that basic research would eventually pay off some time in the distant future came under increasing scrutiny from cost-conscious management. Companies laid off tens of thousands of workers, including scientists and engineers, shut down or sold off entire divisions, and insisted that the survivors turn a consistent profit. They closed central research laboratories and moved researchers back into operating divisions where they could contribute directly to the design of new products. Even a nonprofit like RAND had to diversify into new fields such as healthcare, infrastructure, and energy and the environment, putting more emphasis on policy rather than scientific expertise. Aerospace companies began looking for players who could be counted on for singles and doubles rather than swinging for the fences, and put them to work in buildings that also played it safe. Young scientists seeking their own field of dreams had to look elsewhere, to firms in emerging technologies that offered bolder architecture and a laboratory lifestyle better suited to generations X and Y, who no longer aspired to the blue-sky suburbs of the Cold Warriors.[71] The avant-garde had become, at least in retrospect, the organization men.

Notes

1. Sharon Ghamari-Tabrizi coined the term to describe Kahn's colleagues at RAND. See Ghamari-Tabrizi, *The Worlds of Herman Kahn: The Intuitive Science of Thermonuclear War* (Cambridge, MA: Harvard University Press, 2005).

2. Richard Florida, *The Rise of the Creative Class* (New York: Basic Books, 2002).

3. David Kaiser, "The Postwar Suburbanization of American Physics," *American Quarterly* 56, no. 4 (2004): 851–888.

4. Roger W. Lotchin, *Fortress California, 1910–1961: From Warfare to Welfare* (New York: Oxford University Press, 1992).

5. *Dr. Strangelove or: How I Learned to Stop Worrying and Love the Bomb* (1964 film), Stanley Kubrick, director and producer, Columbia Pictures.

6. Fred Kaplan, *The Wizards of Armageddon* (New York: Simon and Schuster, 1983), 56. See also Martin Collins, *Cold War Laboratory: RAND, the Air Force, and the American State, 1945–1950* (Washington, DC: Smithsonian Institution Scholarly Press, 2002).

7. John McDonald, "The War of Wits," *Fortune* 43 (March 1951), 144.

8. John D. Williams, *An Overview of RAND*, D-10053 (Santa Monica, CA: RAND Corporation, 1962).

9. John D. Williams, "Note to a Workaholic," RAND Corporation Archives (Santa Monica, CA), John Williams Papers, Box 1, f.6.

10. Quoted in Michael Kubo, "Constructing the Cold War Environment: The Strategic Architecture of RAND" (M.Arch thesis, Harvard University Graduate School of Design, 2009), 62.

11. Ibid.

12. Ibid.

13. Specht, "Some Prejudices about the Size and Character of RAND," M-1274, February 2, 1965, RAND Corporation Archives (Santa Monica, CA), Robert D. Specht Papers, Box 6, Folder 5.

14. Alex Abella, *Soldiers of Reason: The RAND Corporation and the Rise of American Empire* (New York: Harcourt, 2008), 94.

15. Timothy Braseth, "Josef van der Kar: Building Architectural Bridges," *Modernism* 14, no. 2 (Summer 2011): 44–53.

16. Leonard McCombe, "A Valuable Batch of Brains," *Life Magazine*, May 11, 1959, 101.

17. Gene Marine, "Think Factory De Luxe," *The Nation*, February 14, 1959, 132.

18. James R. Drake, "Wind Surfing—A New Concept in Sailing," P-4076 (Santa Monica, CA: RAND Corporation, 1969), https://www.rand.org/pubs/papers/P4076.html (accessed January 16, 2018).

19. Covering the program in detail: Stephanie Young, "'Would Your Questions Spoil My Answers?': Arts and Technology at the RAND Corporation," in *Where Minds and Matters Meet: Technology and California and the West*, ed. Volker Janssen, 293–320 (Berkeley: University of California Press, 2012).

20. Ibid., 304–306.

21. *Selections from the Norton Collection at RAND* (Santa Monica, CA: RAND, 1991).

22. Fran Seegull, "Selections from the Norton Collection at RAND," 1991, RAND Corporation Archives (Santa Monica, CA).

23. Kaplan, *The Wizards of Armageddon*, and Abella, *Soldiers of Reason*, include good chapters on Ellsberg's work at RAND.

24. Jennifer Light, *From Warfare to Welfare: Defense Intellectuals and Urban Problems in Cold War America* (Cambridge, MA: MIT Press, 2003).

25. Virginia Campbell, "How RAND Invented the Postwar World," *American Heritage of Invention and Technology* (Summer 2004): 50–59.

26. Abella, *Soldiers of Reason*, 305.

27. Robert Genter, "With Great Power Comes Great Responsibility: Cold War Culture and the Birth of Marvel Comics," *Journal of Popular Culture* 40 (2007): 953–978.

28. Simon Ramo, *The Business of Science: Winning and Losing in the High-Tech Age* (New York: Hill and Wang, 1988); Warren Mathews, "The Personality and Fundamental Management Orientation of the Hughes Aircraft Company," internal document, courtesy of Arthur Chester. Mathews came to Hughes Aircraft in 1949 and went on to become vice-president, responsible for the electronics division.

29. D. Kenneth Richardson, *Hughes after Howard: The Story of Hughes Aircraft Company* (Santa Barbara, CA: Sea Hill Press, 2011), 44–45.

30. L. A. Hyland, *Call Me Pat: The Autobiography of the Man Howard Hughes Chose to Lead Hughes Aircraft* (Virginia Beach, VA: Donning Company, 1993).

31. "Harold Rosen Oral History," *Engineering and Technology History Wiki*, January 15, 2015, http://ethw.org/Oral-History:Harold_Rosen (accessed March 1, 2016).

32. "The Development of Synchronous Satellites," March 9, 1965, Hughes Electronics Collection, MS 2004-10, Special Collections, UNLV Libraries, University of Nevada, Las Vegas.

33. Eugene Moehring, "The Sahara Hotel: Las Vegas' Jewel in the Desert," in Karin Jaschke and Silke Otsch, eds. *Stripping Las Vegas: A Contextual Review of Casino Resort Architecture* (London: Verso, 2004), 14–15; Dennis McLellan, "Ragnar Qvale, 86; Actor, Led Architectural Firm," *Los Angeles Times*, October 2, 2001, http://articles.latimes.com/2001/oct/02/local/me-52287 (accessed March 1, 2016).

34. The Allen Curve, named for MIT researcher Thomas J. Allen, sets 150 feet as the critical distance for weekly communication among colleagues.

35. Joan Bromberg, "George F. Smith Oral History," February 5, 1985, American Institute of Physics (College Park, MD), 19.

36. Andrew Hamilton, "Shangri-La for Deep Thinkers," *Westways* 54 (February 1962): 4–6.

37. Julius Shulman and Juergen Nogai, *Malibu: A Century of Living by the Sea* (New York: Harry Abrams Publishers, 2005).

38. Theodore Maiman, *The Laser Odyssey* (Fairfield, CA: Laser Press, 2001).

39. "Recollections of Richard Feynman by Charles South," *artchester.net,* August 7, 2015, http://artchester.net/2015/04/richard-feynman-personal-impressions/ (accessed March 1, 2016).

40. W. Patrick McCray, *The Visioneers: How a Group of Elite Scientists Pursued Space Colonies, Nano-technologies, and a Limitless Future* (Princeton, NJ: Princeton University Press, 2012).

41. Author correspondence with Arthur Chester, December 16, 2015.

42. "Program for European Press," Monday, March 8, 1965, Hughes Electronics Collection, MS 2004-10, Special Collections, UNLV Libraries, University of Nevada, Las Vegas.

43. Richard Sanders Allen, *The Northrop Story, 1929–1939* (New York: Orion, 1990).

44. E. T. Wooldridge, *Winged Wonders: The Story of the Flying Wings* (Washington, DC: Smithsonian Institution Press, 1983).

45. "Here's a 7-Question Quiz to Help You Decide on Your Future," *The Michigan Technic*, December 1959, 14.

46. "The Peninsula Welcomes Nortronics to Beautiful Palos Verdes Research Park," 1961, Local History Center, Nortronics vertical file, Palos Verdes Library District (PVLD).

47. Victor Gruen and Larry Smith, *Shopping Towns USA: The Planning of Shopping Centers* (New York: Reinhold Publishing, 1960).

48. "Palos Verdes Research Park: Making Science Feel at Home in California," Southwest Museum of Engineering, Communications and Computation, July 1961, http://www.smecc.org/palos_verdes_research_park.htm (accessed March 7, 2016).

49. "Palos Verdes Research Park," 1961, Local History Center, Nortronics vertical file, PVLD.

50. Al Johns, "Palos Verdes Peninsula Stirs Amidst New Growth," *Los Angeles Times*, February 21, 1960, I1.

51. "Gallery: SeaView, Palos Verdes, California," *Paul Revere Williams: American Architect*, October 23 1960, http://www.paulrwilliamsproject.org/gallery/seaview-palos-verdes-california/ (accessed April 24, 2017).

52. Chet Lewis to Project Group, June 1, 1960, Charles Luckman Papers (CLP), 2006.65, Series 3, Box 6, f. 9 Loyola Marymount University Special Collections.

53. Don Peterson to Project Group, August 12, 1959, CLP.

54. "Environment for Research," CLP, undated, architect's presentation brochure.

55. "Northrop's 'Park' Attracts Top Talent," February 14, 1976, Local History Center, Nortronics vertical file, PVLD.

56. "The Peninsula Welcomes Nortronics to Beautiful Palos Verdes Research Park," Nortronics vertical file, PVLD.

57. Betty Lukas, "What Nortronics Does," *Rolling Hills Herald*, 1962, Nortronics vertical file, PVLD.

58. "Northrop Works on Skylab Project," May 20, 1971, Nortronics vertical file, PVLD.

59. Iao Katagiri, "RAND Headquarters Project: Goals, Objectives and Design Guidelines," ca. 2003, RAND Corporation Archives (Santa Monica, CA).

60. Ibid.

61. Sarah Palmer, ed., *Architecture at Work: DMJM Design Los Angeles* (New York: Edizioni Press, 2004), 118–139.

62. Richardson, *Hughes after Howard*, 415–438, details the decline and fall of Hughes Aircraft under GM.

63. Compiled from HRL employee comments on *Glassdoor.com*, https://www.glassdoor.com/Reviews/HRL-Reviews-E140549_P2.htm (accessed November 6, 2015).

64. Roger Vincent and Alex Pham, "New Media Takes Old Hughes Space," *Los Angeles Times,* February 24, 2012, http://articles.latimes.com/2012/feb/24/business/la-fi-hughes-youtube-20120224 (accessed March 6, 2016).

65. Andrea Chang and Peter Jamison, "Playa Vista Turning into Silicon Valley South as Tech Firms Move In," *Los Angeles Times*, January 17, 2015, http://www.latimes.com/business/la-fi-silicon-valley-south-20150118-story.html (accessed April 24, 2017).

66. Shawn Langlois, "Southern California Takes on Silicon Valley in Battle for Tech Supremacy," *New York Post*, July 16, 2015, https://nypost.com/2015/07/16/southern-california-takes-on-silicon-valley-in-battle-for-tech-supremecy/ (accessed March 6, 2016).

67. "Notes on Report of Working Relationship with South Bay State College and Palos Verdes," 1963, CSU Dominguez Hills, Special Collections, Master Planning/Site Selection Collection, Box 4, f., Site Selection, Palos Verdes.

68. "Northrop Sells RHE Facilities," 1979, Nortronics vertical file, PVLD.

69. "Northrop May Sell RHE Land," *Daily Breeze*, March 20, 1991, Nortronics vertical file, PVLD.

70. Ilene Lelchuk, "Building a Future: Northrop Center Being Razed," August 30, 1996, Nortronics vertical file, PVLD.

71. David Beers, *Blue Sky Dream: A Memoir of America's Fall from Grace* (New York: Doubleday, 1996).

5   The Siberian Carnivalesque: Novosibirsk Science City

Ksenia Tatarchenko

In May 1967, a banner above the entrance of the Novosibirsk State University (NGU), a leading Russian university located in Siberia, announced the beginning of a special event. Depicting a trumpeter blowing into his instrument, it bore a slogan proclaiming the advent of the "Freedom of Freedoms." The youthful crowd that had gathered under the banner eventually formed a procession that began to spread out into the streets of the Novosibirsk Scientific Center, known locally as *Akademgorodok*. The streets were the epitome of Nikita Khrushchev's promise of affluence; modern prefabricated houses and public transportation vehicles lined the Akademgorodok's main artery, now home to an energetic procession of monks, gypsies, jesters, ladies garbed in evening gowns, half-naked prehistoric figures, and cross-dressing sailors and musketeers. The immediate implication of the "Freedom of Freedoms," as inscribed overhead, was the ability to assume a new identity.

The masked characters were the students of the NGU campus that, between its inauguration in 1959 to the close of the 1960s, became lauded as the most prestigious school in the region.[1] The University was conceived as a critical element of the new science city rising in the Siberian forest, one that aimed to create a clear distinction between itself and the infamous "closed" towns of the nuclear program.[2] The campus was the embodiment of Akademgorodok's tripartite agenda, later formulated as the motto: "science–cadres–industry."[3] The cavalcade of masked youths embodied the promise of this Socialist scientific organization, responsible both for the production of experts and for the fulfillment of the ambitious long-term goals of the party-state to contribute to world-class fundamental research. In the process Siberia would be developed and brought into the twentieth century.

The tension between the futuristic aspiration of the science city, enshrined in its architectural forms, and the archaic form of festivities enacted on its premises in May 1967, is fundamental to this chapter's analytical focus. In *Rabelais and His World* ([1965] 1984), Mikhail Bakhtin famously exposed the creative and transgressive nature of the

Figure 5.1
Carnival banner "Freedom of Freedoms" by Iu. I. Kononenko, 1967. Photoarchive of the Siberian Branch of the Russian Academy of Sciences.

Figure 5.2
Carnival procession, 1967. Photoarchive of the Siberian Branch of the Russian Academy of Sciences.

medieval carnival.[4] In rupturing hierarchies and their anchoring in time and space, scientific and technological innovation, like a carnival, is a moment containing the potentiality of another "disruptive" world. It is the bodily, collective, and creative aspect of Bakhtin's notion of "carnivalesque" that will inform this chapter's view of Akademgorodok as a space for Socialist innovation and knowledge production. The chapter will reconstruct the forms of education and leisure at the NGU, focusing on the formative exposure of Siberian youths to Socialist science and lifestyle. Further, it will argue that the overlap between the two spheres was not an accidental feature specific to the Akademgorodok, but a structural characteristic of the Soviet scientific community. In other words, while the city scale of Akademgorodok was predicated on the conceptual and spatial configuration of the science city, its material manifestation belongs to a broader phenomenon, also documented in more recent works on late Soviet science.[5]

NGU—its campus, curriculum, and, most of all, the leisure activities of the student body—represents a particularly promising case for interrogating the relationship between the built environment, professionalization, and the lifestyle of Soviet "Big Science." This latter term, popularized by Alvin Weinberg in reference to American mass-laboratories of the post–World War II and Cold War eras, was co-opted by the founders of Akademgorodok to emphasize their efforts and the size of the Siberian territory.[6] In terms of the laboratory lifestyle focus of this volume, this chapter challenges the declinist narratives of previous English-language accounts of the Soviet science city. The historian Paul Josephson, famous for his works on Soviet-era science, depicts Akademgorodok as a "New Atlantis" made possible by its remoteness from Moscow's influence.[7] For the architecture historian Alexander D'Hooghe, Akademgorodok is primarily a cybernetic descendant of the traditional garden city.[8] Both narratives locate the utopian element within the realm of ideas—the early modern sources for scientific ethos and the nineteenth-century socialist movement, respectively—and, ultimately, represent the science city as a utopia corrupted by Soviet reality. This chapter uniquely breaks with that stereotype, both conceptually and methodologically, by introducing a different set of historical sources. It draws on official discourse, student recollections, and photographic evidence to consider utopia as moments of "lived experience."[9] The notion of the carnivalesque facilitates a situated understanding of the personal, scientific, and political aspirations behind the Akademgorodok project, and offers insights into the relationship between the inhabited spaces, social interactions, and knowledge-making particular to it.

This chapter builds on the arguments that the Akademgorodok is best understood as an embodiment of the post-Stalinist relations between power and experts; a model and display at the national and international scale.[10] The term "post-Stalinist" invokes the past as a crucial component of the social and cultural fabric of late Socialism. Approaching Akademgorodok's history via the biography of the city's founding father, mathematician Mikhail A. Lavrentiev, has made it possible to elucidate the networks and power relations permeating the Siberian Branch of the Soviet Academy of Sciences. For instance, the status of the science city as a model and showcase was predicated on the peculiar personal relationship between Lavrentiev and Nikita Khrushchev, first formed in the 1940s during their work in the Ukraine. Lavrentiev's past, as manifest in his age (he was fifty-seven at the time of creation of the city), also affected the nature of his authority among the inhabitants of the new community. Among the circle of his Siberian students and associates, Lavrentiev was known as *ded*—the "grandfather."[11] This nickname expressed a key characteristic of Akademgorodok: the average inhabitant was youthful, many of them under the age of thirty.[12]

Though the science city was new and its residents were young in the 1960s as a result of the targeted hiring policy of the city's founding fathers, this also reflected the national demographic of the time; almost half the population of the USSR's 235 million citizens were under the age of twenty-seven.[13] As individuals, NGU students of the 1960s were representative of the social composition of Soviet society. As a group, their experiences and subcultures offer crucial insight into the new scientific community's agenda and practices, as well as its values guiding professional and private lives. Although any comprehensive discussion of dissent and conformity in Akademgorodok is beyond the scope of this chapter, its chronological framing aims to unsettle the post-1989 tendency to conflate any "fooling about" with political subversion, and to highlight the continuities of practices connecting work and leisure across the late Soviet period.[14]

Setting the Stage: Tomorrow's Progress and Yesterday's Values

The official opening ceremony of NGU's first building was held on April 19, 1963. The institution's first rector, I. N. Vekua, gave a short but important speech to mark the occasion. Holding a symbolic key, he announced: "This is a key from Science! This key shall never close anything at all. Let it have a unique property—to open!"[15] For the city leaders, the act of *opening* was tied to the idea of scientific *discovery*. Critically, in Russian, the two words are formed from the same root, *otkryvat'* and *otkrytie*. In his memoirs, Lavrentiev stressed that there were no useless discoveries, describing the task of science as the discovery of new natural phenomena and their explanations through the creation of theories.[16] Openness was also a reference to the main conceptual innovation of NGU: a pedagogy not enclosed in its own didactics, but embedded in the very infrastructure of the science city. "Novosibirsk University is not only a new institution of higher learning," stressed Vekua when introducing the NGU to readers of the newspaper *Pravda*, "it is also a university of a new type."[17] The standardized features of the building itself were celebrated as characteristic of this typology: NGU was to be literally "diffused" (*rastvoren*) throughout the city.[18] In his memoirs, Lavrentiev described the building with pride:

Our University is unusual. First of all, its building is much smaller than that of traditional universities (the construction cost us only 4 million rubles instead of the usual 40 million). How was this possible? First and foremost, there is not a multitude of laboratories—students do not work with the teaching instruments and models but in real laboratories of the academic institutes. Here, we do not need as many lecture halls—a majority of seminars and electives take place at the institutes. Finally, the University does not have to provide office space to its chair heads and faculty members—they already have them at their main place of work.[19]

In reality, investments in human capital had a far greater role in defining university life than the cost savings of diminishing NGU's footprint. The faceless building became home to handpicked young talent able to be taught at an individual level. The official rhetoric stressed the radical novelty of the approach, with its mechanisms of selective recruitment and individualized learning, emphasis on mathematics and physics, and the concept of early integration into research activities. Such rhetoric veiled the origins of the NGU's pedagogy in an older tradition of polytechnic education. The NGU was a novel reworking of ideas borrowed from an existing model that Akademgorodok's founding fathers, M. A. Lavrentiev, S. A. Khristianovich, and S. L. Sobolev, had already pioneered at the Moscow Physical Technical Institute, better known as "Fiztekh." Fiztekh was an experimental institution for training a new type of specialist for the postwar military-industrial complex. Thus, NGU's roots encompassed technocratic imagination and attempts at institutionalization, from revolutionary Paris to Leningrad to Stalinist Moscow and the Khrushchev Thaw-era Siberian science city.[20]

Similar to their Fiztekh peers, the Siberian students of NGU were treated as an elite group, capable of autonomous work with the most advanced materials and tasked with a workload that bordered on the impossible. All recollections, including those of the students that transferred from St. Petersburg and Moscow, emphasize the incredible difficulty of their studies. There was no introductory or general physics courses and instead, students of all specialties had to take a theoretical physics class designed by Gersh Budker. One of his most infamous quips stated: "I won't teach you physics, but life. You can read Landau on your own."[21] Dark humor in the popular student song of the period highlighted the transcendent aspect of entering NGU, pronouncing death as the only way out: "You shall stop waiting for my return / Here, I forget even the name of my mother / Death, death, death/ is the only escape … / And there is no rest for NGU students."[22] They had little choice but to try to accomplish the feats expected from them, or leave. Many had to take the exams twice, and the dropout rates were high.

There was one major difference between Fiztekh and the NGU, however. Unlike the experimental institution, admission to the NGU was not predicated on a thorough screening process. On the contrary, many of the NGU's students came from the provinces and Siberia, where family histories marked by forced exile were commonplace and brought with them potential difficulties entering other prestigious universities. For example, the Soviet Republic of Kazakhstan, the involuntary home of a community of Russian-Germans (among other displaced ethnic groups), was traditionally well represented at the NGU. The university, with its emphasis on future employment in academia and industrial research, represented an opportunity for upward mobility.

The new students to NGU were a receptive audience for the official discourse of the city's leadership, which stressed the novelty and unique character of the university. Constantly reminded about their status as a vanguard of Socialist construction, the Siberian science city's "workers of tomorrow" looked toward the future with confidence. They were eager to believe that Soviet science, and Akademgorodok in particular, would "influence the scientific progress in the world," as one could read on the pages of the local newspaper.[23] Though their daily lives were highly structured by academic schedules, they were complemented by shared leisure activities that contributed a great deal to feelings of community and belonging. If Akademgorodok was to embody the Socialist promise of plenty, a balance between production and pleasure was necessary for individual creative powers to flourish.

Extra-curriculum: The Heralds of Freedom

The 1967 carnival was an example of the students' creativity in action. Visual arts, music, and performance combined in a variety of forms of artistic expression to reflect upon Akademgorodok's burgeoning cultural life. The kaleidoscopic nature of the photographic record in the archive of the Siberian Branch of the Academy of Sciences highlights the interdependence between individual experiences and the collective dynamic of the event; a tacit consensus that the carnival was a moment worth remembering. Unlike recollections influenced by retrospective reflections and later events, the photographic evidence on the carnival is a valuable source providing an additional perspective on Akademgorodok's famous social life and prominent social clubs.

Post-1991 accounts frequently mention the city's leisure activities as an expression of the "independent" spirit of the science city—where "independent" could just as easily be read as "protest." This particular connotation also appears in the often-cited expression that depicts the city as "a kingdom of fearless (nepuganyk) fools," associating the transgression implied by "foolishness" and impolite behavior with an impending repression, where fear is instilled by the authorities.[24] This part of the chapter demonstrates that such an association of foolishness or frivolity with repression is misleading, and that mass events such as the carnival served a dual function that was sanctioned by the local authorities and university administrators. On one hand, the carnival belonged to mechanisms of professional identity formation, and filled a function in direct correlation with NGU's main agenda. On the other hand, it was part of a larger set of social and cultural activities, existing as a mechanism of social cohesion in an urban community populated by newcomers. The carnival heralded togetherness, announced the freedoms of association, but did not imply freedom from the party-state regime.

The focal site of the event had an improvised stage built on the stairs leading to the entrance with amplifiers and loudspeakers and decorated by Iu. I. Kononenko. Dancing was an essential part of carnival festivities. Later the stage turned into a podium where the contest for the esteemed title of "Carnival King and Queen" was held. It was the point of departure for a lively procession. Composed of floats devised by individual faculties, the parade was led by the village carriage, decorated with a carpet and drawn by students. The cart was occupied by S. T. Beliaev, the rector of the university from 1965 to 1978. He shared it with the two masters of the carnival, who together managed the crowd from their loftier position via megaphones. The carnival obeyed the commands of K. M. Skobeev, an economics student dressed in an extravagant hat, and the jester A. B. Khutoretskii, who stayed at the NGU graduate school upon obtaining a diploma from the mathematics department in 1966.[25]

The images show that mockery made up the general spirit of the crowd. Some costumes, such as the infamous Ku Klux Klan gowns or representations of jobless Americans from the 1930s, were overtly political. However, mocking Western symbols was neither a direct order of the party nor a sign of the reverence toward the Soviet ideology among the future elites. The crowd was dancing the twist, and recollections reveal a fascination and familiarity with Western music and culture among the student body. Moreover, the crudeness of the ideological discourse—imposing a model behavior as it appeared on the pages of the university's English-language textbook—prompted derision. One of the most ridiculed stories recalled by students was about a certain Rosa Shafigulina, a "simple Soviet girl" who offered her eyes to alleviate the suffering of an old American communist leader, an act described as a "typical example of the spirit of the Soviet youth." "How many Rosa Shafigulinas are there in the Soviet Union?" asked the text. "Millions and millions."[26]

This story struck the imagination of all those involved not only because of its absurdity, but also because of its preoccupation with the body; a theme typically silenced in both official and individual accounts beyond the conventional celebration of physical activities and sports.[27] Sexuality—the invisible, private aspect of student lives—was also at play during the carnival, and costumes show that the male-female dichotomy carried more importance than geopolitical opposition between the East and West. Cross-dressing was a frequent occurrence, with female students dressed as sailors and musketeers, and a number of male students posing as female gypsies and ladies in gowns. Gender transgression manifested the life-defining questions that students faced. In Soviet universities, it was commonplace for students to marry and raise children, a pattern that no doubt influenced the gendered job pyramid typical of late Soviet society.[28]

The youths' preoccupation with sexual experience was universal, and so it naturally became a satirical theme. Carrying the words "I search for love as a cat in May" across his back, one carnival participant was not alluding to the madness of the English "March hare," but to the sexual drive and promiscuity evoked by the Russian expression "March tomcat."

This student, who would later describe his own masked identity as that of a "perverted angel," was Vladimir Sabinin. Classmates remembered Sabinin for his extraordinary skills in chess, and as a frequent contributor to the sardonic newsletter produced in one of the dormitories. In the paper, Sabinin assumed a double authorial voice and identity; he wrote satirical reports under the name of Evgenii Mosampilov, and erotic poetry as the female poet Lubov' Shal'naia (*Stray Love*). Unlike the serious literary meetings held among Akademgorodok's young poets as they rediscovered and venerated the heritage of the Russian "Silver Age," Sabinin and his friends were simply irreverent jesters.[29]

Their newsletter was first called *Uni-TASS* (playing on the Russian word for WC [water closet, or bathroom/toilet] and the name of the Soviet news agency). It was later transformed into the handmade magazine *PromeTASS*, this time a reference to one of the official student newspapers, *Prometheus*. The unserious "toilet" humor of this literary venture reflected some of the "lowly" interests of the elite student body: dubious lexicon, excessive alcohol consumption, and sex. A few memoirs also testify that samizdat copies of the *Kama Sutra* and Freud were popular reading materials circulating among the *PromeTASS* audience in the dormitories.[30] During the Carnival's march across the city, the contributors of *PromeTASS* carried a banner displaying the words: "Oh, Cover Your Pale Legs!" The banner's reference is to a single line absurdist poem by Valery Bryusov from 1894. The form Bryusov invented is known as a monostich. Although we cannot know whether the educated citizens of the new science city were amused or appalled by monostich, the larger context of the group activities performed by the students suggests it was meant to stir controversy.

The carnival celebrated a unique moment regarding the "freedom of freedoms," but generally, liberties allowed on campus had to be negotiated with administrative bodies. In the absence of statistics regarding expulsion for disciplinary violations, memoirs help reconstruct the university student lifestyle. They depict dormitories as spaces of liberation from parental control; drinking, broken windows, and jokes involving condoms are explicitly described in the memoirs, and a few hints regarding suicide or sexual and ethnic abuse are also alluded to. According to V. A. Mindolin, who assumed the post of "party secretary" during his time at NGU, during a heated discussion about

Figure 5.3
Vladimir Sabinin as "Perverted Angel," 1967. Photoarchive of the Siberian Branch of the Russian
Academy of Sciences.

Figure 5.4
"Oh, Cover Your Pale Legs!," 1967. Photoarchive of the Siberian Branch of the Russian Academy of Sciences.

a series of unspecified incidents in student dorms, the rector opted to accommodate some liberties of student lifestyle.

The choice of language and policy by Beliaev was emblematic of the time and place. Himself a graduate of the Fiztekh, Beliaev would successfully combine an administrative career with scientific research, and became known for his contributions to quantum many-body theory. As the rector observed: "we shall always oscillate between the anarchy and the caserne. We won't admit anarchy. But we do not want to have a caserne. What follows? We need to learn how to live in an oscillating regime."[31] As an administrator, Beliaev was known for his reliance on student self-governance. In 1965, the "academic councils" were elected to work alongside the Komsomol councils and serve as intermediate bodies for solving practical administrative problems related to managing student internships and the postgraduate practice of job distribution.[32] From the point of view of many student participants, the carnival was first and foremost an occasion to have fun; however, for Beliaev's administration, the temporal disorder and loss of authority catalyzed collective identity formation.

Thanks to their connections within Moscow's scientific milieu and the Moscow State University (MGU), the power of festivals for socio-professional bonding were no doubt familiar to Beliaev and other Akademgorodok leaders. Starting in 1959, the annual celebrations for the "day of physics" attracted several thousand participants at the MGU. Additionally, a small cadre of guest "stars" were known to make appearances; Lev Landau in 1960, Niels Bohr in 1961, and cosmonaut Herman Titov in 1963. Although distinct for its tradition of amateur opera, most notably the comic "Archimedes," the Moscow festivities were structurally alike to the NGU's Siberian carnival. A staircase was transformed into a stage where humorous recollections on university life took place, followed shortly by a costume competition. Pyrotechnical effects and gadgets were part of the show and, of course, marching and dancing. As the number of physicists steadily grew, these celebrations fostered the formation of a distinct disciplinary and cultural identity. In Moscow, as in Siberia, participation in collective festivities helped integrate young scientists into a professional community ethos.[33]

Its MGU predecessor notwithstanding, the NGU carnival was also deeply rooted in the local culture of mass celebration and amateur theater. For instance, only a few months prior to the carnival's procession down Akademgorodok's streets, another holiday was celebrated in the city's public areas—"Maslenitsa," or the "Welcome of Spring," similar in vein to Mardi Gras. Maslenitsa's festivities and street theater performances involved many the same participants as the carnival, and sported visuals produced by the very same artist.

The role of such local events for social cohesion is similar to those that can be observed in other newly created scientific communities. The social fabric of these communities depended on the integration of newcomers, not dissimilar to those "instant" American settlements, most notably Los Alamos.[34]

While shaped by the needs of the community, cultural resources, and practices of the period, the NGU carnival and its occupation of the city's streets was also one point in a long tradition of Soviet political culture. Even if Kononenko's artwork belonged to a tolerated underground movement, and was not wholly representative of the official aesthetic of Socialist Realism, the form of the street mass festivity itself was not subversive of the dominant culture. The student carnival procession was a play version of the official Soviet parades, and a form of reincarnation for the revolutionary mass festivities that recreated mythologized narratives. Biannual mass processions, formed by representatives of local factories and institutions, were a hallmark of Labor Day,

Figure 5.5
"Maslenitsa," street theater performance, March 1967. Photoarchive of the Siberian Branch of the
Russian Academy of Sciences.

and the commemoration of the October Revolution on November 7 in Soviet cities
throughout the period.[35]

Shafigulina's eyes, Symbolist's legs, and Freud's nose—this composite portrait, made
of cultural references current among the selected youths of the model socialist city
does not look like the official version of Soviet culture, but nor is it a depreciative
image. Situating the 1967 carnival within a tradition of mass festivities is an important
condition for appreciating the relative importance of 1968, a year often described as a
watershed moment in the history of the Siberian science city, and that of late Socialism
more generally.

Beyond 1968: Carnivalesque Uninterrupted and Transformed

Two major events had already shaken the science city's community by the time of
the infamous events of August 1968. In February, scientists and NGU faculty staff
from Akademgorodok signed the "letter of 46," addressed to Soviet party leaders to
protest the violation of juridical procedures during the trial of the four dissidents.[36]

In juxtaposition to the seriousness of these proceedings, the famed Akademgorodok social club, "Under the Integral," held the national bard festival in March. Featuring a gala performance by Alexandre Galich, the festival is now remembered as a mark of Akademgorodok's boldness. This was to be Galich's first and last public performance before forced exile in 1974.[37]

The student body did not escape the turmoil of 1968. Signs of protest began to appear, scrawled on the city walls, first in the spring, then again in August. The graffiti was the initiative of a small circle of political activists among NGU students.[38] While only very few actually partook in the rebellious act, the prospect of repression—the expulsion of the wrongdoers—sparked a mass reaction from the student body. Beliaev had to face the improvised gathering within the dorms, and, while promising to do all in his power to lighten consequences, spoke out against the actions with an emphasis on the responsibility that came with the privilege of belonging to the model community. Ultimately, he sought to highlight the intellectual and material privileges that were given to intelligentsia of the USSR, and in particular at Akademgorodok, in comparison to the conditions facing peasants and workers.[39] Official leaders, Beliaev included, articulated their commitment to what they believed were the value of the intelligentsia within society, and implied that the protesters had abandoned these values.[40]

The events experienced by the Siberian community in 1968 retrospectively add political undertones to the previous year's festivities. One encounters a frequent idea that the social life within the science city changed dramatically, the change being often attributed to the closing up of the club Under the Integral, which held its last meeting to celebrate the 1970 New Year with the masked "Ball of Imbeciles."[41] Yet student carnivals were not forbidden, and several more took place through 1971. In 1969, the physics department even received assistance for its participation in the parade from a military helicopter that carried its banner high above the city. Moreover, the same forms of creative expression—student clubs, newspapers, humorous performances, and amateur theater—continued and would eventually bring the Novosibirsk University teams to the national stage on the televised and widely popular student improvisation contest, the Club of Merry and Resourceful.[42] Discerning continuity, without denying the importance of changes that marked Akademgorodok's collective memory, provides an explanation for the connections between different forms of social life within the science city. These connections were embodied in both the festivities that structured the communal life of the scientific institutions—the improvisations, dressed parties, and satirical poetry—as well as in the reemergence of the Physics Day at NGU during the early 1980s.

Figure 5.6
Interweek, students carry a dummy representing the "rotting West," 1980. Photoarchive of the Siberian Branch of the Russian Academy of Sciences.

NGU's new major mass festivity, beyond the carnival of the late 1960s, was shaped by a distinct form of political discourse. In the 1970s and 1980s, the NGU became famous for its week of International Solidarity, or "Interweek." Held every May, it gathered activists, international participants, creative collectives, and student artists all in the one spot. In line with the official political discourse of the era, the Interweek was also an occasion for large-scale festivities featuring song contests, posters, and satirical theater performances, this time with an approved subject of mockery—the imperialistic ambitions of the "rotting" West.[43]

Very much like Kononenko's poster, the iconography of the Interweek celebrated the advent of "freedom," however, its imagery depicted an autonomy obtained through liberation from capitalist aggression, and fueled by a fear of war. For those participating, freedom did not take the form of an individual assuming a new identity; rather, it was a freedom discovered in collective action. However, even under Lenin's gaze, the large crowd kept gathering in the same yard to laugh and sing together.

Figure 5.7
Interweek, closing ceremony, 1983. Photoarchive of the Siberian Branch of the Russian Academy
of Sciences.

In the 1990s, the festivities were stripped of their ideological function and evolved
into a gathering fulfilling two seemingly opposite functions: a forum for discussing
higher learning at the time of economic crisis, and a large-scale music festival. The
persistence of mass events, even in this insipid form, highlighted the enduring need for
social integration and cohesion in the ever-youthful student body.[44]

Conclusions

Akademgorodok was constructed as an embodiment of late Socialist urbanism, characterized by close links between work, residence, and social facilities. When the
Soviet party proclaimed science a productive force of socialism, however, it did not
spare funds to build a community representative of the power that planned science
and an attractive Soviet lifestyle could manifest. Modest individual apartments dis-
tributed to the scientific workers, and a handful of town houses allocated to elites,
combined with the well-supplied commercial center to create much better living conditions than those experienced by most of the Soviet population. Excellent schools and

numerous recreational facilities, in addition to beautiful natural surrounds, contributed to Akademgorodok's appeal, an attraction that persists up to this day. For several
decades the NGU was the best entry point to join this model community.

Akademgorodok's founding fathers conceived the university in such a way as to
ensure the durability and self-reproduction of the Siberian experiment. From an enormous construction site, the new university managed to acquire a student body and
a temporary headquarters as early as 1959. The repetitive and standardized character of the NGU's buildings became an object of pride, explained as the materialization of the efficiency of the experimental educational concept behind the new school.
Its main principles, such as careful student selection, emphasis on mathematics and
physics, and laboratory work, were not as novel as claimed, and in fact belonged to
the long tradition of polytechnic education. But the NGU successfully modernized,
enacted, and popularized these ideas, both across the country and internationally. It
added leisure pastimes and community building to exacting scholarship and prominent faculty. Social and cultural activities, such as the 1967 carnival, allowed individual
self-expression and consolidated Akademgorodok's social fabric.

According to Lavrentiev, "A person who wants to become a scientist has to learn
how to work during [their] leisure time."[45] This comment did not refer to the frequent
and expected practice of "after hours" laboratory work—that went without saying.
Instead, very much like *Monday Begins on Saturday*, the famous title of the 1964 scientific fairytale by the Strugatsky brothers, it pointed to the overwhelming nature of
scientific labor as a habit of the mind, a creative process, and group membership that
ultimately defines the individual.[46] This vision is crucial for understanding how the
Akademgorodok's spaces were inhabited, and the 1967 carnival remains the most visually striking manifestation of the overlap between private and professional spheres.
During the event, proclamations of disciplinary identification and private bodily concerns were mocked, exposed, and engraved onto the streets of Akademgorodok. Such
was the social importance of this event that the large-scale student festivities persisted
across the watershed year of 1968, albeit with ideological realignment. The celebration
of the "freedom of freedoms" turned into the celebration of liberation from imperialist
oppression, but remained a collective celebration nevertheless. The bodily, collective,
and creative aspect of "carnivalesque" remained integral to Akademgorodok's lifestyle
across the decades.

Notes

1. The new university became a competitor of the Tomsk State University, founded in 1878 as
the Imperial Siberian University, and a series of newer institutions, including Irkutsk State Uni-

versity (1918); Omsk Machine-building Institute (1942); and Krasnoyarsk Polytechnic Institute (1956). In the rapidly growing Novosibirsk, several institutions for higher learning were created to supply a workforce for the industrialization of the 1930s and following the evacuation during World War II, including the Siberian Building Institute (1930); Railway Engineering Institute (1932); Medical Institute (1935); Pedagogical Institute (1935); Agricultural Institute (1936); Elec-trotechnical Institute (1950); Institute of Water Transport Engineering (1951); Electrotechnical Communication Institute (1953); and Institute of Soviet Cooperative Commerce (1956).

2. For a comparison of lifestyles in the cities of the Soviet and American nuclear programs, see: Kate Brown, *Plutopia: Nuclear Families, Atomic Cities, and the Great Soviet and American Plutonium Disasters* (Oxford: Oxford University Press, 2013).

3. Zamira Ibragimova and Natalia Pritvits, *Treugol'nik Lavrentieva* (Moskva: Sovetskaia Rossiia, 1989).

4. Mikhail Bakhtin, *Rabelais and His World* (Bloomington: Indiana University Press, 1984).

5. For a description of a playful local subculture at the campus of the Ukrainian Institute of Cybernetics, see Benjamin Peters, *How Not to Network a Nation, The Uneasy History of the Soviet Internet* (Cambridge, MA: MIT Press, 2016), 126–136. On infrastructures sustaining leisure activi-ties in the "atomic" city of Obninsk, see: Zinaida Vasilyeva, "Der unauffällige Staat: Die Infra-struktur der Amateurverbände in der Zeit des Spätsozialismus," in *Hochkultur für das Volk? Literatur, Kunst und Musik in der Sowjetunion aus kulturgeschichtlicher Perspektive*, ed. Igor Narskij (Berlin: De Gruyter/Oldenbourg, 2018), 213–234.

6. Compare Alvin M. Weinberg, "Impact of Large-Scale Science on the United States," *Science* 134 (1961): 161–164; and Mikhail A. Lavrentiev, "Razvitie nauki v Sibiri i na Dal'nem Vostoke," *Vest-nik AN SSSR* 12 (1957): 3–7. See also Gennadii Pospelov, "V nauchnom tsentre Sibiri," *Nash sovre-mennik 3* (1958): 319–321; and Gennadii Pospelov, "Bol'shaia nauka v Sibiri," *Sibirskie Ogni* 12 (1959): 159–170.

7. Paul Josephson, *New Atlantis Revisited: Akademgorodok, The Siberian City of Science* (Princeton, NJ: Princeton University Press, 1997).

8. Alexander D'Hooghe, "Science Towns as Fragments of a New Civilization: The Soviet Develop-ment of Siberia," *Interdisciplinary Science Reviews* 31, no. 2 (2006): 135–148.

9. "Lived experience" in connection to "utopia" here does not diminish the tension between "good" and "nonexistent" produced by homophonous *eutopia* and *utopia* in the direction where "experience" entails corruption and dystopia. Instead, it localizes historical actors as they imag-ine and make efforts to create a plausible version of the future. This approach is influenced by the discussion of utopia in Michael Gordin, Helen Tilley, and Gyan Prakash, *Utopia/Dystopia: Condi-tions of Historical Possibility* (Princeton, NJ: Princeton University Press, 2010).

10. Ksenia Tatarchenko, "Calculating a Showcase: Mikhail Lavrentiev, the Politics of Expertise, and the International Life of the Siberian Science-City," *Historical Studies in the Natural Sciences* 5 (2016): 592–632.

11. Ravil Garipov, "Po stsenariiu Lavrentieva," in *Vek Lavrentieva*, ed. Natalia Pritvits, V. D. Ermikov, and Zamira Ibragimova (Novosibirsk: Zdatel'stvo SO RAN, 2000), 181–285.

12. Tamara Ilatovskaia, "V poiskakh 'sumasshedshei' idei," *Smena* 840, May 1962.

13. Statistics from *Life* (November 1967).

14. For a recent work approaching dissent as a narrative, a form of self-understanding and a set of experiences closely tied to daily life, see: Jonathan Bolton, *Worlds of Dissent: Charter 77, The Plastic People of the Universe, and Czech Culture under Communism* (Cambridge, MA: Harvard University Press, 2012). Also see Benjamin Nathans, "Talking Fish: On Soviet Dissident Memoirs," *Journal of Modern History* 87 (2015): 579–614.

15. Ilia Vekua quoted in *NGU: Universitetskaia zhizn*, November 16, 2004. The symbolic key is preserved at the NGU history museum.

16. Mikhail. A. Lavrentiev, "Opyty zhizni," reprinted in *Vek Lavrentiev*, 350. A large body of literature in the sociology of science is devoted to the notion of "discovery" in scientific discourse; a classical reference is Bruno Latour and Steve Woolgar, *Laboratory Life: The Construction of Scientific Facts* (Princeton, NJ: Princeton University Press, 1986).

17. Ilia Vekua, "Universitet novogo tipa," *Pravda*, June 19, 1959.

18. Ivan Meshkov, "Tri istochnika i tri sostavnye chasti," in *Vek Lavrentieva*, 324–326.

19. Lavrentiev, "Opyty zhizni," 170. It is interesting to compare a university embedded within a science city to American campuses described as "cities in microcosm" and utopian social visions of America. See the classical account by Paul Vernable Turner, *Campus: An American Planning Tradition* (Cambridge, MA: MIT Press, 1984).

20. The literature on École Polytechnique is large. For a classical account see: Terry Shinn, *L'École Polytechnique: 1794–1914* (Paris: Presses de la Fondation nationale des sciences politiques, 1980). For a description of the distinct professional identity described as "engineer scientists," see Ivor Grattan-Guiness, "The 'Ingeneur Savant,' 1800–1830: A Neglected Figure in the History of French Mathematics and Science," *Science in Context* 6, no. 2 (1993): 405–433. On the Leningrad community, see Paul Josephson, *Physics and Politics in Revolutionary Russia* (Berkeley: University of California Press, 1991). On the history of Fiztekh and its connection to the educational experiments in Leningrad, see Nikolai Karlov, *Povest' drevnikh vremen, ili predistoriia Fiztekha* (Moscow: MTsGO MFTI, 2005).

21. "Landau" is a reference to the multivolume course book on theoretical physics famous for its difficulty. It was co-written by Lev Landau and Evgenii Lifshitz, and first published in 1940 and reedited in 1958. Lev Landau and Evgenii Lifshitz, *Mekhanika* (Moscow: Fizmatgiz, 1958).

22. Svetlana Ivaniia, "NGU—po volnam moei pamiati," http://gold.vixpo.nsu.ru/default.aspx?db =museum_1959&int=VIEW&el=3855&templ=VIEW (accessed April 1, 2016).

23. Richard Courant, *Za nauku v Sibiri* 33, August 1963.

24. A variation on the set phrase "'Krai nepuganykh idiotov.' Samoe vremia pugnut,'" which loosely translates to "a country of idiots who know no fear. It is the perfect timing to scare them." Attributed to the notebooks maintained by the famous Soviet journalist and writer Iliia Il'f from 1925 to 1937 and partially published in 1939, 1957, 1961. Also see Iliia Il'f, *Iz zapisnykh knizhek* (Leningrad: Khudozhnik RSFSR, 1966).

25. On Khutoretskii, see his biography on the NGU website, *Novosibirsk State University,* https:// nsu.ru/ef/kafedra_ef_primeneniya_matematicheskih_metodov_v_ekonomike/hutoreckii _aleksandr_borisovich (accessed on April 1, 2016).

26. Jokes about Rosa Shafigulina were the identity marker of NGU graduates for many years. See the online discussion at http://forum.nsu.ru/viewtopic.php?p=34901 (accessed on April 1, 2016).

27. Sports and tourism were extremely popular in Akademgorodok and deserve a detailed exploration that is beyond the limits of this chapter. On tourist culture of the 1960s, see Anne E. Gorsuch and Diane P. Koenker, *Turizm: The Russian and East European Tourist under Capitalism and Socialism* (Ithaca, NY: Cornell University Press, 2006).

28. See Ksenia Tatarchenko, "'The Computer Does Not Believe in Tears': Soviet Programming, Professionalization and the Gendering of Authority," *Kritika* 19, no. 4 (2017): 709–739.

29. On the literature circles in Akademgorodok of the 1960s, see Gennadii Prawkevich and Tatiana Ianushevich, *Zelenoe vino: Literaturnyi Akademgorodok shestidesiatykh* (Novosibirsk: Svin'in i Synov'ia, 2009).

30. Recollection available at Vladimir Sabinin's website, http://ffngu64.ucoz.ru/Uvlechenija/Prometass.pdf (accessed April 1, 2016).

31. Vladimir Mindolin, "Nostal'giia," in *I zabyt' po-prezhnemu nel'zia,* ed. Natalia Pritvits and Svetlana Rozhnova, 169–174 (Novosibirsk: IEOPP SO RAN, 2007).

32. Komsomol, or the Communist Youth League, was the key organization structuring student life. Compare with Komsomol practices described in Benjamin Tromly, *Making the Soviet Intelligentsia: Universities and Intellectual Life under Stalin and Khrushchev* (Cambridge: University Press, 2013), 159–186. On the combination of formal and informal mechanisms in Komsomol, see Aleksei Yurchak, *Everything Was Forever, Until: The Last Soviet Generation (*Princeton, NJ: Princeton University Press, 2006), especially chapter 3, 77–125.

33. Iu. V. Gaponov, "Traditsii 'fizicheskogo iskusstva' v rosiiskom fizicheskom soobchshestve 50-90kh godov," *Voprosy istorii estestvoznaniia i tekhniki* 4 (2003): 165–178.

34. Jon Hunner, *Inventing Los Alamos: The Growth of an Atomic Community* (Norman: University of Oklahoma Press, 2007).

35. James von Geldern, *Bolshevik Festivals, 1917–1920* (Berkeley: University of California Press, 1993).

36. For a detailed, document-based account, see Ivan Kuznetsov, *Pis'mo soroka shesti. Dokumental'noe izdanie* (Novosibirsk: Klio, 2007).

37. On Galich and bard movement, see Amy Garey, "Aleksandr Galich: Performance and the Politics of the Everyday," *Limina: A Journal of Historical and Cultural Studies* 17 (2011): 1–13. At the time of the festival, Galich was not yet formally "forbidden," and would be excluded from the official Soviet organizations after the event.

38. An English-language account mentioning some of those activists can be read in Raissa Berg, *Acquired Traits: Memoirs of a Geneticist from the Soviet Union* (New York: Viking Penguin, 1988).

39. Beliaev's speech is reported in an interview with Sergei Krasil'nikov at http://gefter.ru/archive/17807?_ut_t=fb (accessed April 1, 2016).

40. Here "intelligentsia" is an actor category and not an analytical term. Defining "intelligentsia" is a major topic in Russian historiography; the "creative" intelligentsia typically draws more scholarly attention than the "technical" meaning. The major work on the late Soviet intelligentsia emphasizing connections to the prerevolution values of "social justice" is Vladislav Zubok, *Zhivago's Children: The Last Russian Intelligentsia* (Cambridge, MA: Belknap Press, 2009).

41. Anatolii Burshtein, "Ot 'goroda Solntsa' k gorodu 'Zero,'" *I zabyt' po-prezhnemu nel'zia*, 127–135.

42. On KVN, see: Christine Elaine Evans, "From Truth to Time: Soviet Central Television, 1957–1985" (PhD thesis, UC Berkeley, 2010), chapter 5. NGU won national contest in 1988, 1991, and 1993.

43. See a 1983 press report: V. Aleksandrov and D. Zolotov, "Tak derzhat'!," *Krokodil* 22, August 1983.

44. On student political activism in Siberia, see Aleksai Borzenkov, *Molodezh´ i politika: Vozmozhnosti i predely studencheskoi samodeiatel´nosti na vostoke Rossii (1961–1991 gg.)*, 3 vols. (Novosibirsk: Novosibirskii Gosudarstvennyi Universitet, 2003).

45. Lavrentiev, "Opyty zhizni," 356.

46. Arkadii Strugatskii and Boris Strugatskii, *Ponedel'nik nachinaetsia v subboty: Skazka dlia nauchnykh rabotnikov mladwego vozrasta* (Moscow: Detskaia literatura, 1965).

6 Scientific Dining

Sandra Kaji-O'Grady

In the popular imagination scientists are secretive loners and awkward geniuses, as caricatured by the fictional Sheldon Cooper in *The Big Bang Theory*, whose bafflement at human interaction and emotion is the television series' comedic engine. Against this stereotype, one of the principle objectives of scientific research organizations is to coax scientists into sociability. Architecture is favored as a vehicle for eliciting gregariousness. Yet providing comfortable spaces for conversation, or lively "streets" and atriums, is rarely sufficient incentive to lure scientists away from the bench. Hence, in the design and operation of scientific research buildings, much emphasis is placed on staff tearooms and restaurants, as well as the food and beverages served in them.[1] This is why the founder of Stanford University's Bio-X, Jim Spudich, advises other research managers to "cancel the labs and build the cafeteria" should they find themselves faced with budget cuts.[2] The historical shift in meals in the workplace, as Altman and Baruch write, is that meals, "once a marginal anecdote to institutional life," are "now a well-considered organizational feature."[3] The lunch break, they argue, should be viewed as fulfilling a role in shaping, maintaining, and negotiating shared meanings, and in interpreting identities that are institutional as well as individual.[4] Meals sit alongside other "soft" phenomena pertinent to organizational effectiveness, while also establishing associations between work and wider societal concerns.

To fully understand scientific dining and its relationship to organizational objectives, we must look at: menus, prices, and gastronomical contents; dining spaces and furniture; catering equipment; chefs, kitchen and wait staff; food and restaurant brands and their meanings; and meal etiquette and traditions. We must also take in the engagement scientists themselves have with their in-house restaurants and cafés, for this is as revealing as the attitudes expressed by scientific managers. With all of this in mind, this chapter is concerned with three aspects of workday dining in the contemporary laboratory building: nutrition, wellness, and care of the self; commensality and workplace culture; and the rhetorical deployment of dining as a signifier of lifestyle. It argues that these three aspects collude to construct the individual scientist as a self-determining, creative, and collegial worker, a fantasy that is at once antidote, means, and distraction from the scientist's increasing exposure to the market forces of a competitive knowledge economy.

Writing about Scientific Dining

The middle section of Jeffrey Eugenides's 2011 novel *The Marriage Plot* is set in the fall and winter of 1982–1983 on the campus of the Cold Spring Harbor Laboratory (CSHL) on Long Island, New York. Eugenides's wife, sculptor Karen Yamauchi, had lived in the nearby town of Cold Spring Harbor for a year, and he relied on her impressions to recreate the scene. This seems like a tenuous connection out of which to construct a convincing backdrop, but so successful is Eugenides in recreating the atmosphere of the campus in the 1980s that Dr Amar Klar "was flooded with 20-year-old memories" when he read the novel.[5] Klar claims Eugenides "got the facts and the setting of the institute right."[6] Klar is the geneticist whose real work on yeast at CSHL was the basis for the research that engages the character of Leonard in the book.

In *The Marriage Plot*, Cold Spring Harbor is reimagined as Pilgrim Lake. Its long-standing director, James Watson, appears as the figure of Dr. Malkiel, and Barbara McClintock is thinly disguised as Diane MacGregor, her 1983 Nobel Prize incorporated in the novel's plot. Leonard wins an internship to Pilgrim Lake and is accompanied there by his girlfriend, Madeleine. Eugenides writes that Madeleine

hadn't expected the limousines ferrying pharmaceutical executives and celebrities in from Logan to eat with Dr. Malkiel in his private dining room. She hadn't expected the *food*, the expensive French wines and breads and olive oils hand-picked by Dr. Malkiel himself. Malkiel raised huge sums of money for the lab, lavishing it on the resident scientists and luring others to visit. It was Malkiel who bought the Cy Twombly painting that hung in the dining hall and who had commissioned the Richard Serra behind the Animal House. ... Almost every night there was a party where people did slightly queer, science-nerd things, such as serving daiquiris in Erlenmeyer flasks or evaporating dishes, or autoclaving clams instead of steaming them.[7]

Watson did, in fact, host private dinners with prospective donors, and in 1994 had a house on the campus built to be "suitable for large-scale entertaining."[8] He boasts about the culinary talents of his wife, Elizabeth, in his memoir *Avoid Boring People* ("consommé Bellevue" topped with horseradish-flavored whipped cream featured at one of these business meals in 1974). The fine dining had at home in "Ballybung," though, did not extend to Watson's subordinates and the staff dining hall. Indeed, Madeleine's favorable impressions are contradicted by an unauthorized biography of

James Watson by Victor McElheny, a science writer employed by the CSHL between 1978 and 1982, the same period "recalled" by Madeleine in the novel. In researching *Watson and DNA: Making a Scientific Revolution* (2003), McElheny interviewed forty-five scientists, many of whom had worked under Watson at the CSHL.[9] They relate an intensely competitive, small-town atmosphere, overseen by the mercurial Watson. Scientist Phil Sharp remembers that "It had all the bad aspects and all the good aspects of a commune."[10] Heiner Westphal comments they all "knew each other too well. They definitely stepped on each other's toes all the time. You would know what was frying in your neighbor's pan at night."[11]

The cold reality of life at CSHL is confirmed in a 1995 review of the food served at its dining hall, published in the *Annals of Improbable Research*, a science humor magazine edited by the founder of the parody Ig Nobel Prize. CSHL's Blackford Hall is lampooned for offering "mediocre food at mediocre prices," and for serving up the remains of the Saturday night lobster banquet as "lobster bisque" on Sunday and "half-price salad" on Monday.[12] An entrée of Shrimp Nuremberg is "chunky," "yellowish," and "somewhat recognizable." The "stoic" décor has worked so well, snipes the reviewer, that it has remained unchanged since construction in 1907.[13] The review is illustrated with a grim black-and-white archival photograph of the empty dining hall from the 1920s. The concrete dining hall was upgraded in 1992 to increase its capacity from 170 to 400, but aesthetically it remains true to its origins. Before the upgrade "the lunch lines in the Laboratory's dining hall sometimes extended not only through the dining room itself, but also out and across Bungtown Road."[14] Until 2010, Blackford Hall was the sole food outlet on a campus, accessible only by car or infrequent shuttle bus from the rail station, and served around 450 lunches daily.

Blackford Hall was not unique in serving disappointment; indeed, the review is one of a series called "Scientific Dining: Reviews of Research Institute Cafeterias," targeting the canteens and cafeterias of the world's eminent research centers. The aroma of the Scripps Research Institute cafeteria in La Jolla, California, for example, is likened to that at a McDonald's, its décor pronounced "bland, solid and clean."[15] Next door to a hospital, the reviewer warns "it is not unusual to find yourself seated next to someone with an IV holder in her arm or electrodes protruding from her head."[16] The canteen staff were observed to have an obsessive interest in the return of the dinner trays. The relatively tasty food and cleanliness at the Whitehead Institute for Biomedical Research in Cambridge, Massachusetts, was considered "a betrayal of the cafeteria standards honored by most of the world's other, older research institutions."[17] The subsidized canteen was, indeed, encountered by scientists globally. Reflecting on the "awful cafeteria" at the Maxwell Ayrton-designed National Institute for Medical Research (NIMR, 1949) in

Figure 6.1
Blackford Hall at the CSHL in 2015. Photograph by the author.

Mill Hill, London, a scientist recently recalls that it "fostered a sort of wartime solidarity."[18] It is this sense of shared suffering, of humor in the face of adversity, that underpins the "Scientific Dining" series.

The culinary distance between the yellowish shrimp dish at Blackford Hall and the handpicked olive oils served by the self-described "happy hostess"[19] Elizabeth Watson at home (or the fictional Malkiel) is the product of real and imagined social hierarchies, as well as different modes of writing. While motivations and readership differ, the novelist, the biographer, the author, and the scientist-cum-restaurant reviewer share an underlying belief that what scientists eat in the workplace, along with where and with whom, is revealing of scientific life. But one must ask: Exactly what is it that is being revealed? Exposition in Eugenides's narrative contextualizes the tensions building between Madeleine and Leonard as he descends into mental illness, but how does an account of scientific dining help us understand the lives of real scientists? What is revealed about the internal social organization of scientific research centers and their relationship to broader society? And what meanings are ascribed by managers and scientists to what and where scientists eat, and with whom they share a meal?

Shaping the Scientist through Nutrition

The menu at Blackford Hall is, today, little changed—a buffet bar heavy on carbohydrates and fried food. It is not just the lack of competition for diner patronage lessening the need for CSHL to improve its dining facilities. The benefits of no teaching or administrative responsibilities and the freedom to pursue inquiry in this elite institution outweigh the inferior food and austere hall. Yet, such fare is now an exception. Researchers from the NIMR in Mill Hill, referred to previously, were transferred to the Francis Crick Institute in 2016 where they now enjoy coffee from "specialty roasters" in a cafeteria that its architects describe as "intentionally small to foster chance encounters."[20] The shift from instant coffee to specialty roasters mirrors broader social change, changes that saw the "Scientific Dining" series peter out by the start of the new millennium for lack of content.

This development reflects the penetration at all levels of society of a gourmet "foodscape," a recently coined term that embraces the places, people, meanings, and material processes related to food and its consumption.[21] Celebrity chefs, food bloggers, televised cooking competitions and demonstrations, the slow food movement, farmer's markets, and artisanal ingredients are all elements of the democratization of luxury food. Even molecular gastronomy has become commonplace—a form of cooking popularized by Heston Blumenthal that, with its syringes, bottled nitrogen gas, scalpel blades, and surgical tweezers, looks a great deal like laboratory science. In 2013, *Science Daily* highlighted recent statistics on the foodie phenomenon in the United States that found that three-quarters of U.S. adults enjoy talking about new or interesting foods. Fifty-three percent of U.S. adults regularly watch cooking shows, and two-thirds purchase specialty foods for everyday home meals.[22] The lifestyle implications of the phenomenon are captured by the denotation "foodie," defined by Johnston and Baumann as "somebody who thinks about food not just as biological sustenance, but also as part of their identity, and a kind of lifestyle."[23] Diet is always a reflection of cultural identity, yet "foodie" is reserved for a particular kind of elite cultural engagement. As Michael Featherstone notes, the dynamic of consumer culture follows the breaking down of sumptuary restrictions to make some luxuries more widely available.[24] Exhibiting that dynamic, gastronomy has expanded and its meanings shifted. Knowing the origins of one's coffee beans and being on a first-name basis with a favorite barista follows the model of *connoisseurship* that in previous eras expressed itself as an ability to recognize and expound upon the genealogy and beauty of a Ming vase.

It is not just a broadening of audience for artisanal and luxury foods that has taken place, however, for food is caught up in a discourse around longevity, health, and fitness that has been referred to as the "wellness revolution." Scientists feed this discourse and their new findings on the relationship between food and health are eagerly taken up in the mainstream press. Thus, the health effects of consuming chocolate, red wine, coffee, processed meats, sugar, and "superfoods" such as kale and goji berries, as determined in the laboratory, are almost constantly in the news. Functional foods such as calcium and bacteria-enriched yogurt, designed by breeding or additives to manage or prevent specific diseases or problems, or enhance mental performance or metabolic fitness, have blurred the boundaries between pharmaceutical products and foods and beverages. Food is medicine, literally and metaphorically. Dieting now aims at empowerment, rather than weight loss. Health is conceived as a "resource for living," and eating well a means to both longevity and extended youthfulness, increasing what has come to be called one's "healthspan," the length of time where one is in optimal health.[25] The shift is captured by the slogan of food and beverage giant Unilever following its corporate rebranding in 2004: "Looking good, feeling good, get more out of your life." A 2010 Nike advertising slogan directed at women vehemently pitches the same idea: "You are entirely up to you. Make your body. Make your life. Make yourself."[26]

In the United States, employers are the second major driving force of the expanding wellness market.[27] The European Network for Workplace Health Promotion has been active in the same arena. Alongside the benefits of healthy food in their canteens, research organizations, like other businesses, are investing in on-site gymnasiums or corporate membership rates to nearby gymnasiums, as well as yoga classes, standing desks and standing meetings, health screenings, influenza vaccination programs, and more. The ambitions are quite unlike the paternalism or "caring capitalism" of the Cadbury brothers who provided canteens at the factory town of Bournville, along with medical and dental treatment and decent housing. At Bournville, workers' canteens were consistent with the Cadburys' belief in the social rights of workers. Today, companies are offering wellness programs to reduce their healthcare costs and employee absenteeism, and to produce workers engaged in an ongoing shaping of identity and body.[28] There is a tacit expectation that employees look after their health and fitness. Thus, the gymnasia and restaurants featuring healthy food options are driven by employers as much as by employees seeking to gain an edge in the competitive vocational environment. Amanda Waring, observing the use of health clubs by high-earning professional workers, describes the development and maintenance of a fit and healthy body as an integral part of a "project of the self" leading to enhanced career opportunities.[29] She notes, too, the competitive nature of the fitness regime and the ways in which ideas drawn from the workplace, of commitment, effort, routine, and reward, are transferred to the fitness club. What looks like leisure is, instead, best understood as an extension of work, not its antithesis. As Derek Wynne observes, in this way "the dominance of work as central to life produces a pattern of leisure practice in which work interests predominate."[30]

The situation has intensified as Carl Cederström and André Spicer make brilliantly clear in their sardonic anatomy of the "wellness syndrome."[31] Pointing to corporate rules that go as far as banning not just smoking from the workplace, but also smokers, the authors show that wellness has become an ideology that demonizes those perceived as not looking after their bodies. Obligatory wellbeing has replaced social welfare, encouraging us all to be mindless athletes in the name of productivity, with the perhaps unintended effect of condemning the poor to a lack of medical support. As David Harvey observes, under capitalism sickness is defined as an inability to work.[32] Frew and McGillivray concur, arguing that health and fitness clubs "oil the desire and dreamscape of physical capital, maintaining an aesthetic masochism and thus keeping the treadmills literally and economically turning."[33]

In this context, leisure pursuits are shaped by work-like concerns, and this applies, too, to the lunch break. Indeed, it is not a "break" from work at all, but an extension of its concerns. This aspect of scientific dining is clearly apparent in the rhetoric around the James H. Clark Center (2003) at Stanford for BioX. Designed by the architects Foster and Partners in collaboration with MBT Architecture, the building includes on its ground floor a full-service cafeteria. On the third floor is a branch of Peet's Coffee that "only sells artisan-crafted coffees."[34] The cafeteria at BioX is called "Nexus," a word that refers to connections and links, and in cell biology, to the membrane involved in intercellular communication and adhesion. It is one of four hundred corporate cafés in the United States owned and managed by Guckenheimer, a contract food service provider established in 1963 by two Stanford students. The Nexus Internet homepage declares that "Healthful dining is an integral part of a wellness program." Furthermore, it states that "our culinary team uses the freshest, seasonal, local, sustainably grown, and organic ingredients we can obtain. Everything we serve is made from scratch— from salad dressings to soup stock. We do not use frozen, processed and canned food, nor do we use trans-fats, additives, preservatives, or excessive added salt. We use nutritional cooking practices such as braising, broiling, steaming, and poaching."[35]

Selling meals is coupled at Nexus with nutritional coaching and counseling for Stanford University's employees and students. The implications are clear. Employees are expected to engage in a program of self-improvement, shaping their bodies and minds in ways that benefit themselves and the organization. Likewise, the Salk Institute for Biological Studies in La Jolla has adopted this doctrine. Its café is one of the CulinArt chain, dedicated to encouraging "our customers to maintain a healthy lifestyle—one that is holistic and incorporates balanced food choices, regular exercise, and an overall attention to living well."[36] The approach is not confined to Californian sites of science. In Switzerland, at Restaurant Cloud on the Basel campus of pharmaceutical corporation Novartis, the focus is on "ecologically-produced and fair-traded produce."[37] Menus note the provenance and organic certification of ingredients, alongside the caloric intake of each meal. The Loft, at the Victorian Comprehensive Cancer Centre in Melbourne, Australia, a research and clinical center opened in 2016, serves "healthy food … freshly prepared for optimal growth, physical vitality and social well-being."[38] The implication, in the context of diners who are mix of scientists, medical professionals, and cancer patients and their visitors, is that those ill with cancer have failed to maintain good nutrition. It's a subtle form of the victim blaming that Susan Sontag railed against in *Illness as Metaphor* (1979).

What one eats also enables scientists as social agents to make a statement about the self, to *perform* a particular vocational identity. It is why Nina Tandon, who we met in chapter 1, feels it necessary to explain her response to an ethical dilemma: "I choose not to eat meat, but do choose to engage in experiments that involve the sacrifice of animals."[39] She explains her vegetarianism as compensation for her "karma footprint."[40] Craig Venter on the other hand, is actively researching genetically modified plants and animals for food, and describes himself as an omnivore, eating a healthy mixed diet with plenty of fish.[41] Both emphasize their self-determination, and their insertion of personal bodily management, within public debates that touch their work. Such dietary choices, although informed by an understanding of their environmental, ethical, or health consequences are, for these two, also consistent with generational, geographical, and familial situations. Tandon's vegetarianism is unsurprising given she is a left-leaning, urban, professional woman in her thirties from New York, but such choices are heightened for scientists as they overlap with vocational identities.[42] The relationship scientists have with food can be especially complex where it intersects with their research interests. Venter, for example, foresees a role for synthetic biology in solving food shortages and the environmental cost of agriculture; be that through meat grown in the laboratory from microbes, or reengineered algae that absorbs carbon dioxide and can be used as feedstock. While in the laboratory pursuing synthetic or what he calls "motherless meat," he makes sure that this research pursuit is not misunderstood as a vegetarian's passion project, stating "I like good steaks, but I like lots of things that aren't sustainable."[43] His after-dinner speech in Turin in 2012 is especially revealing of the preoccupations he brings to the table. Venter tells fellow diners gathered in the ornate nineteenth-century premises of the Michelin one-star Ristorante del Cambio: "We've just finished this fantastic meal with a rough count of about twenty or thirty different species whose DNA we all consumed and, despite how well it was cooked, a few billion microbes on the side. And the microbes in your gut are metabolized in the chemicals that you absorbed, and you all have about fifty chemicals right now circulating through your brains from bacterial metabolites of what we just ate."[44]

How his audience reacted to this is not recorded, but it reminds us that the scientist sees every meal as a lesson in predation and biological evolution, and an illustration of the action of chemical processes on bodies that are eaten, and bodies that have eaten. Dining is also a psychic event, an ethical conundrum, a point in the global flow of production and consumption, a test of one's knowledge of etiquette, and a moment for asserting one's identity, taste, and social class. Alongside this, it is shared with others and, thus, a cultural and social occasion. Insert all of this into the workplace and scientific dining is fraught with the possibility of catastrophic social failure and paralyzing ethical conundrums, while equally holding the potential to transform selves and organizations.

Socializing the Collective

For many older scientists, breaking for lunch is an indulgence (one is reminded of the movie *Wall Street*'s Gordon Gecko declaring that "Lunch is for wimps").[45] A broader change in business management discourse, however, has substituted collegiality, creativity, and innovation for older fixations with efficiency and productivity. By the late 1990s, Taylorism, with its focus on imposing a fixed pattern of movement from above, had been supplanted by team-based structures aimed at "constant improvisation in work organization and the unobtrusive orchestration of employee values."[46] In place of compliant bodies, the new high-involvement workplace sought supple minds—commitment supplanted control. In scientific research organizations, this change is expressed as a belief that breakthrough discoveries will be accelerated by increasing the opportunities for scientists to be creative, collegial, interdisciplinary, and connected. This hope has placed intense pressure on the midday meal and the coffee break as a tool of subtle social control. For some managers, this potential is best exploited away from the laboratory as a special occasion. Joerg Schaefer, director of the Cosmogenic Dating Group at Columbia University's Lamont-Doherty Earth Observatory, for example, claims, "I make it clear that I expect everyone who works here to have fun. We have lunch together once a month, off campus."[47] For most managers, though, regular communal dining on site is the ideal vehicle to shape informal socializing. Hence, the changes that have taken place in the laboratory canteen—now ubiquitously upgraded to a restaurant, café or "culinary center"—are targeted at the health of the collective enterprise as much as individual well-being.

In a promotional article on the first-year anniversary of the completion of the Clark Center at Stanford, the university emphasized the positive relationship between its in-house café and scientific progress:

The sharing of ideas doesn't just take place in the labs and offices but in the center's third-floor Peet's café and its ground-floor cafeteria, LINX, which has long communal tables to promote conversation between strangers. "I run into people at Peet's all the time," remarked Karlene Cimprich, an assistant professor of molecular pharmacology and yearlong resident of the Clark Center. In fact, her regular 9:30 a.m. coffee run has led to a joint venture with Aaron Straight, an assistant professor of biochemistry, who goes over to Clark for coffee from his lab in the Beckman Center. The two of them struck up a conversation one morning while waiting in line for their caffeine fix and discovered that they both were trying to solve the same puzzle.[48]

Stanford University's media release for the celebration of the Clark Center's first decade reiterates, "like the kitchen at a great party, NeXus, the Clark's culinary center, is where everyone ends up."[49]

At the Max Planck Institute for Molecular Cell Biology and Genetics (MPI-MCBG) in Dresden, Germany, the construction of a cafeteria was central to management's vision for the new institution and the design of its building. The MPI-MCBG is one of twenty institutes created after the unification of Germany in 1990 in the five New States and former East Berlin. It moved into a purpose-built center designed by Finnish architects Mikko Heikkinen and Markku Komonen, in cooperation with laboratory specialists HENN, at the end of January 2001. On a long and narrow site, several kilometers from the historical center of Dresden, the laboratory building consists of six wet laboratories, two on each floor, connected by a central hall that opens to the full height of the building. The entrance, library, auditorium, espresso bar, and restaurant on the ground floor are pivotal to the institute's operations.[50] Wieland B. Huttner, one of the five directors relates that in addition to recruiting staff and getting the laboratory infrastructure in place, they also had to "ensure the presence of an operational canteen. The last aim was of particular concern to at least one of the directors, Marino Zerial, who took charge, and I gladly confess that I do not miss the Mensa of Heidelberg University at all!"[51]

The restaurant is a double-height room on the ground floor that seats roughly one-hundred people around long, eight-person tables. One wall of floor-to-ceiling glazing looks out onto a garden terrace, while the inner glazed wall faces the entrance lobby.

Figure 6.2
The staff restaurant at the MPI-MCBG in Dresden in 2014. Photograph by the author.

It is an attractive, bright space serving around 350 main dishes each day. As the lunch break is sacrosanct in Germany, the restaurant opens only between noon and 2:00 p.m. and is designed for a lunch-time rush. There are no kitchens or staff tea facilities outside of the canteen and café, compelling scientists to use the subsidized communal facilities. The MPI-MCBG provides the canteen space, kitchen, kitchen equipment, and operating costs to a private operator without charging any rent or fees. They also pay for repair, maintenance, and replacement of kitchen equipment.

The MPI-MCBG's canteen maintains a long German tradition, one that persisted longer in the former East Germany in which Dresden is a major industrial center. Beginning in the nineteenth century with industrialization, workers and public authorities at first resisted workplace canteens as a threat to the coherence of the family. For the capitalist owners of factories, canteens were seen as a means of controlling the workforce, "improving eating habits and lowering alcohol consumption."[52] Canteens gained importance under the Nazi regime as a symbol of the coming *Volksgemeinschaft* and then, during World War II, grew in response to food shortages and increased numbers of women working outside of the home. Canteen use peaked at 77 percent of all workers in the East during 1988. Tatiana Thelen explains that there was a legal obligation in the German Democratic Republic to provide the workforce with food and community eating was seen as superior to individualistic consumption.[53] Canteens came to be seen as progressive, a worker's right, and a means of female emancipation. MPI-MCBG's canteen with its strong employee participation reflects this history. While it is operated privately, it has a canteen committee whose members meet quarterly to discuss improvements. Staff are surveyed regularly for their feedback on food and quality and the canteen committee reviews these responses together with the canteen operator.

Zerial's efforts, and those of the other directors, seemed to have paid off. In a survey by *The Scientist* in 2009, the MPI-MCBG was voted the "Best Place to Work" for postdocs. But are the scientists using lunch to further their research, as management hopes? When Holden Thorp and Buck Goldstein visited the Clark Center they describe how a walk around the Nexus restaurant at lunchtime "evidences an extraordinary set of conversations. Chemists talking to doctors. Math geeks with laptops pointing at simulated models of virtual lungs. Engineers and physicists looking at pictures of medical devices."[54] The impassioned talk overheard by this author at the MPI-MCBG restaurant was not, however, about science. It concerned the assault of the celebrity chef Nigella Lawson by her (then) husband outside Scott's restaurant in Mayfair, an event that had been reported across the globe less than a month earlier. Indeed, this author's eavesdropping on informal conversations between scientists at numerous social spaces in laboratories has yielded little more than the ordinary banter of work colleagues about the weather, current affairs, and weekend sports, and, of course, the personal and professional fortunes of colleagues and other scientists.

Portrayals of laboratory cafés as hotbeds of intellectual exchange are likely shaped by marketing objectives. Indeed, the inflated rhetoric around informal socializing in laboratory buildings is as idealized as the scene Eugenides imagined for the Cold Spring Harbor Laboratory. Discovery is typically an incremental process engaging many scientists over a long period. Rarely, if ever, is it a light bulb moment with a near-stranger over a macchiato. The mythology of "accidental" scientific discovery coming out of serendipitous conversations persists because there aren't post-occupancy studies that consider the performance of laboratory architecture against their socialization ambitions, over and above the obvious effects of co-location. The impact of the canteen and other design artifacts in the workplace is, as Brandt and Lonsway noted in chapter 2, a constructed fact repeated so often and with such confidence that it has become naturalized. The lack of empirical evidence in this regard is intensified by the fact that research on food and commensality is similarly thin. As Sturdy, Schwarz, and Spicer noted, "with a few rare exceptions, there has been very little research on this topic [meals and eating] in organizational studies."[55] What was written about food and eating by anthropologists in the early- to mid-twentieth century dealt mostly with feasts, cannibalism, food taboos, fasting and sacrifice, the hunting, gathering and growing, and preparation of food by nonurban peoples. More recent studies of commensality focus on households and families rather than workplaces and colleagues. An American study found that one in four people ate lunch with coworkers, and postulated that this was due to "ritualized work group obligations, comradeship and sociability, or relaxation and stress management."[56] The study probed no further, which is to say that we do not know for sure what scientists talk about over lunch or whether luncheon conversations, in fact, lead to new collaborations and research ventures that would have not otherwise happened. But we do know what scientific managers are saying about those conversations, and we can see the material investments being made in the construction of dining facilities. So, while we cannot measure their impact on discovery, we can get a sense of how valuable dining facilities have become in creating organizational identities in a competitive sector.

Rhetorical Uses of Luxury Consumption

Despite its earlier support by capitalist factory owners and Germany's fascists, the canteen is associated with socialist and communist regimes, and with dour institutional settings such as mess halls, student refectories, and boarding schools sharing a style of interior décor to be revived only in nostalgic or ironic terms. It is not associated with fine dining and choice; hence, the effort made to secure the "restaurant" status of on-site dining facilities in both Europe and the United States, and the focus on communicating lifestyle through a distinct architecture. At the campus of the Novartis Institutes for Biological Research in Basel, the European regime of a common and compulsory lunch hour is preserved, but a single canteen is replaced by a selection of restaurants, including a Starbucks. In emulation of a town center, these restaurants occupy the ground floors of laboratory and office buildings facing the main pedestrianized street of the campus. With the exception of a building by Frank Gehry, the architecture at Novartis is subdued, conforming to a prescriptive master plan that dictates height, material, and form. The restaurants, in contrast, are thematically styled and distinctive. Employees and guests can eat Italian food off white china and linen tablecloths at small tables in Osterio Dodici, or Thai meals at Cha Cha in a building designed in the minimalist fashion by David Chipperfield. Here, diners gather around sturdy, but finely

Figure 6.3
Cha Cha's restaurant on the ground floor of the David Chipperfield–designed laboratory, Fabrik-
strasse 22, Novartis campus in Basel in 2013. Photograph by the author.

crafted communal timber tables (the ironic canteen interior), seated upon bench seats,
with the theater of the kitchen in full view. There are burgers and salads at Feelgood,
and artisanal food at Cloud, giving employees the illusion of choice and an urbane
lifestyle. Indeed, one of the restaurants is called "Choice." A permanently parked ice
cream van (in a "street" with no traffic that ends abruptly in a high boundary wall) and
a grocery with bottles of wine displayed on oak barrels in its glass shopfront complete
the scenographic recreation of a village intended by the master plan for the revitalization of the campus.

Novartis has the benefit of being a very large organization with around five thousand employees and up to two thousand visitors on any given working day. For smaller
organizations, higher-quality facilities are made possible by making them open and
attractive to the broader public. This strategy has the benefit of linking the scientific
community to the neighborhoods and cities that host them. At the Center for the

Unknown in Lisbon, Portugal, for example, the on-site restaurant, Darwin's Café, occupies its own building on the waterfront campus and is open to the public for lunch and dinner. While Charles Correa designed the laboratory campus as a set of sculptural and curvaceous white forms around a limestone paved plaza, the interior of Darwin's is a contrasting and eclectic mix of modern architecture and superficial references to old world science. It features rich red fabrics, dark timber wainscoting, walls of faux books, vintage drawings of animal species, portraits of scientists, and quotes about life rendered in neon lights. The interior was designed by the "architectural department" of the Lanidor clothing franchise and is headed by Antonio Runa, who runs Lanidor's in-house cafés. The lunch menu features heady dishes such as "Stuffed Pig's Cheeks with Brussels Sprouts, Bacon and Apple" as well as "Octopus Tentacles over Crumbed Sweet Potato and Shallots." There are over eighty different bottled wines and champagnes available, and an extensive cocktail list. It's a long way from CSHL's dining hall. Darwin's Café aims to make the Center for the Unknown a destination and draw additional revenue from its waterfront site. Its co-location with the laboratories, however, also brings glamor and gloss to the institution, from which it can parlay its other revenue raiser—cancer treatment in a spa-like setting for those benefiting from private medical insurance.

Like the Center for the Unknown, the most prominent public space in the Northwest Corner Building (2009) at Columbia University, New York, is its café. The café occupies a double-height volume, wrapped in full height glazing, at the corner of a building designed as the gateway to the Morningside campus from the intersection of 120th and Broadway. It is the gateway to the gateway. Devised by Rafael Moneo to house physics, chemistry, and biology laboratories and classrooms, it is the organization of the escalators, stairs, and café that is the building's central design feature. Those entering the campus from this direction pass alongside the café, taking the escalator up to the campus plaza, with the café below in full view. It is one of ten locations for Joe the Art of Coffee, the New York-based single-origin coffee roaster. Founded in 2003 in Manhattan's West Village, Joe's presence in this most visible and busy space in the Northwest Corner Building affords the scientists based here association with the hippest cultural elite. Joe's boasts catering to Saatchi and Saatchi, the Alvin Ailey dance troupe, the art dealers Phillips Depury, the artist Jeff Koons, Microsoft, the Natural History Museum, Coach, Guerlain, Guess, and Lucky Brand Jeans.[57] The soundtrack is R&B and hip-hop and the coffees are pricey, but with free Wi-Fi the café is always full. Here, the intended audience for the science that takes place above is the knowing postgraduate student, the postdoctoral fellow, and the promising young scientist.

Figure 6.4
Joe the Art of Coffee on the entrance floor of the Northwest Corner Building, Columbia University, New York, in 2015. Photograph by the author.

At the other end of Manhattan, the Alexandria Center for Life Science (2011), just above the East Village, has two on-site restaurants, Riverpark and Little River. The restaurants stand apart from the laboratory buildings in their own glazed pavilion on a "unique garden plaza with romantic East River views."[58] The *New York Times* describes the restaurant as "shimmery" and "pretty" with its pin-lighted dining room.[59] The restaurants, we are told in Alexandria's promotional literature, are "curated" by the celebrity chef Tom Colicchio. Colicchio doesn't actually cook here, but the culinary style of Andrew Smith, the executive chef, "directly aligns with Colicchio's eco-conscious initiatives, such as utilizing every part of an ingredient and composting unused portions, to minimize food waste."[60] The menu is seasonal and local, drawing on the adjacent Riverpark Farm, one of the largest urban farming models in New York City. The farm, comprising vegetables, herbs, and fruit trees planted in seven thousand cloth-lined milk crates, occupies the site of the stalled construction of the third tower of the Alexandria

Figure 6.5
Little River restaurant at the Alexandria Center for Life Science, New York, in 2017. Photograph by Russell Hughes.

Center. The menu comprises dishes such as "Tuscan Kale Salad with Poached Apples, Walnuts, Pecorino, and Apple-Balsamic Vinaigrette" and "Sautéed Royal Sea Bass with Heirloom Carrots, Wheat Berries, Pomegranate, Spiced Pumpkin Seeds." Simpler and slightly cheaper eat-in or take-out lunch can be bought at Little River, but here, too, the ingredients for sandwiches signal luxury and health, with arugula pesto, buffalo mozzarella, Portobello mushrooms, and charred red onions as fillings. Again, there is a target audience in the science sector, this time it is pharmaceutical executives, angel investors, and mid-career scientists. The strategy is working. By 2015, Alexandria's tenants included Pfizer Inc., Eli Lilly and Company, Roche Holding AG, Cellectis, Intra Cellular Therapies, Kallyope, Kyras Therapeutics, Lycera Corp, and TARA Biosystems.

Another speculative life-sciences laboratory building development, i3 in San Diego's University Towne Center, includes a restaurant, herb garden, and dining terrace. It was completed in 2016 by BioMed Realty Trust for USD 189 million, and immediately leased to the genomics company, Illumina, for ten years. The promotional video, produced by digital agency Studio 216, depicts a youthful clientele mingling in a café with blackboard coffee menu. Others lounge on a sunny terrace. In the foreground, market style timber crates bear baskets of improbably large and colorful ripe fruit. A soothing voiceover declares, "the campus invites employees and guests to socialize, exercise, and dine without ever having to leave."[61]

The scientific enterprise must operate as an incitement to a way of life, not a directive. One can leave, but shouldn't want to, or be seen to want to. The scientist's desire for discovery, self-fulfillment, and professional recognition smooths the way for their co-option into organizational ambitions around collaboration and social coherence, and the pursuit of individual wellness and health. In upscale laboratory dining all the contradictory exhortations of neoliberal capitalism converge: "treat yourself," "connect creatively," "be entrepreneurial," "live to your full potential," "exercise your individuality," "eat well, live longer," "work hard, play hard." While developers, realtors, managers, and architects conspire to construct settings that fulfill the functional and social aspirations of the research undertaking such that scientists have no need to leave, at the contractual level scientists are transient. When they leave, it may not be their choice. The MPI-MCBG, for example, despite the image reinforced by its communal canteen, is a precarious place for scientific careers. Contracts, be it for postdocs or group leaders (but not the five directors) are untenured and limited to five years—this is argued in terms of maintaining innovation and flattening hierarchies. Hyman explains that limited appointments mean "they will take with them their experience of a modern research facility and have a revivifying effect on other institutions."[62] What looks like the continuation of a socialist tradition with its communal tables, subsidized meals, and designated hour, is adapted within a neoliberal landscape of labor. It is not solidarity that is being cultivated here, rather the social skills of individual entrepreneurship.

Returning to the improvement in food quality described at the beginning of this chapter, the puzzle is that the food available to scientists has improved, as have the settings in which they eat, while their employment conditions have deteriorated. There is an inverse correlation between the facilities provided to the scientists on site—the restaurants, gymnasiums, and spaces for informal socialization—and their material security. Along with healthier bodies and lunch-time conversations, management seeks the positive effect and improved morale in employees that food can deliver. Analiese Richard and Daromir Rudnyckyj argue that such subtle and coercive, seemingly munificent managerial practices permit the influencing of conduct and conditioning of behavioral outcomes.[63] Accordingly, the experience of eating tasty, healthy, restaurant-quality food with colleagues elicits positive feelings of belonging in the workplace that are critical at a time of economic transformation.[64] Scientific dining and its impact is not therefore a side effect of the economic changes that have produced the lifestyle phenomenon in science: it is central to its transformation. Sentiment may be notionally excluded from the experimental method, but it is very much in play in the facilities that scaffold the laboratory proper. The design of dining is as much a part of the laboratory lifestyle as its architecture.

Yet, as the "Scientific Dining" series affirmed, the skepticism and humor that scientists bring to the workplace—the very creativity and independence that capitalism requires and seeks to exploit—threatens to undo the efforts—sometimes heavy-handed—of managers and bioscience realtors. As one scientist comments, "You want interaction? Serve beer. More scientific interaction happens over frosty pints than almost anywhere else … but I have not yet seen a pub in a biotech/pharma."[65]

Notes

1. Holden Thorp and Buck Goldstein, *Engines of Innovation: The Entrepreneurial University in the Twenty-First Century* (Chapel Hill: The University of North Carolina Press, 2010), 79.

2. Ibid., *Engines of Innovation: The Entrepreneurial University in the Twenty-First Century* (Chapel Hill: The University of North Carolina Press, 2010), 79.

3. Yochanan Altman and Yehuda Baruch, "The Organizational Lunch," *Culture and Organization* 16, no. 2 (2010): 128.

4. Ibid., 128.

5. Gina Kolata, "The Scientist Was a Figment, But His Work Was Real," *New York Times,* February 13, 2012.

6. Nancy Parrish, "Genetics Research Discovered in a Bestseller," *Poster*, NCI at Frederick, June 7, 2012, https://ncifrederick.cancer.gov/about/theposter/content/genetics-research-discovered-best seller (accessed March 1, 2017).

7. Jeffrey Eugenides, *The Marriage Plot* (New York: Farrar, Straus and Giroux, 2011), 173–174.

8. Elizabeth Watson, *Grounds for Knowledge: A Guide to Cold Spring Harbor Laboratory's Landscapes and Buildings* (Cold Spring Harbor, NY: Cold Spring Harbor Laboratory Press, 2008), 127.

9. Victor K. McElheny, *Watson and DNA: Making a Scientific Revolution* (New York: Perseus/Wiley, 2003), xii.

10. Ibid., 199.

11. Ibid.

12. Karen Hopkin, "Scientific Dining: Reviews of Research Institute Cafeterias. Blackford Hall," *Annals of Improbable Research*, 1995, https://www.improbable.com/airchives/paperair/volume1/v1i3/cold.htm (accessed February 20, 2017).

13. Ibid.

14. Elizabeth L. Watson, *Houses for Science: A Pictorial History of Cold Spring Harbor Laboratory* (Cold Spring Harbor, NY: Cold Spring Harbor Laboratory Press, 1991), 278.

15. Stephen Drew, "Scientific Dining: Reviews of Research Institute Cafeterias. The Scripps Research Institute Cafeteria," *Annals of Improbable Research* 2, October 1996, https://www.improbable.com/airchives/paperair/volume2/v2i5/scripps.htm(accessed February 10, 2017).

16. Ibid.

17. Stephen Drew, "Scientific Dining: Reviews of Research Institute Cafeterias. The Cafeteria at the Whitehead Institute for Biomedical Research," *Annals of Improbable Research* 2, December 1996, https://www.improbable.com/airchives/paperair/volume2/v2i6/white.htm (accessed February 10, 2017).

18. Pez, comment on Derek Lowe, "Fancy Building, Fancy Science," *In the Pipeline,* AAAS *Science*, February 4, 2009, http://blogs.sciencemag.org/pipeline/archives/2009/02/04/fancy_building_fancy_science (accessed February 18, 2017).

19. Watson, *Grounds for Knowledge*, 208.

20. "Nature Explores Design of Francis Crick Institute, London's Collaborative Super Lab," HOK, February 15, 2016, http://www.hok.com/about/news/2016/02/15/nature-explores-design-of-francis-crick-institute-londons-collaborative-super-lab/ (accessed April 5, 2017).

21. See Greg Richards, "Evolving Gastronomic Experiences: From Food to Foodies to Foodscapes," *Journal of Gastronomy and Tourism* 1 (2015): 5–17.

22. Institute of Food Technologists (IFT), "'Foodie' Movement Gains Momentum," *ScienceDaily* https://www.sciencedaily.com/releases/2013/02/130213152124.htm (accessed April 5, 2017).

23. Josée Johnston and Shyon Baumann, *Foodie: Democracy and Distinction in the Gourmet Foodscape* (New York: Routledge, 2015), 1.

24. Michael Featherstone, "Luxury, Consumer Culture and Sumptuary Dynamics," *Luxury* 1 (2014): 47–69.

25. The term "healthspan" came into popular use in the 1990s.

26. "Nike Make Yourself Team," *Nike News*, https://news.nike.com/nike-womens/news/nike-make-yourself-team (accessed May 24, 2015).

27. Ilona Kickbusch and Lea Payne, "Twenty-First Century Health Promotion: The Public Health Revolution Meets the Wellness Revolution," *Health Promotion International* 18 (2003): 275–278.

28. D. B. Moskowitz, "The Bucks behind the Wellness Boom," *Business and Health* 17, no. 2 (1999): 43.

29. Amanda Waring, "Health Club Use and 'Lifestyle': Exploring the Boundaries between Work and Leisure," *Leisure Studies* 27 (2008): 295–309.

30. Derek Wynne, *Leisure, Lifestyle and the New Middle Class: A Case Study* (London: Routledge, 1998), 26.

31. Carl Cederström and André Spicer, *The Wellness Syndrome* (Cambridge, UK: Polity Press, 2015).

32. David Harvey, "The Body as Accumulation Strategy," *Spaces of Hope* (Berkeley: University of California Press, 2000), 106.

33. Matthew Frew and David McGillivray, "Health Clubs and Body Politics: Aesthetics and the Quest for Physical Capital," *Leisure Studies* 24, no. 2 (2005): 161.

34. "Peet's Coffee and Tea," Bio-X, https://stanford.edu/group/biox/clark/peets.html (accessed April 2, 2016).

35. "Nexus Café," Nexus, http://nexusdining.com/clients/nexus/fss/fss.nsf (accessed April 2, 2016).

36. "Wellness," Culinart Group, http://www.culinartgroup.com/wellness/ (accessed April 28, 2017).

37. "About Us," *Die Gastronomiegruppe*, https://zfv.ch/en/microsites/novartis-restaurant-cloud/about-us (accessed April 3, 2016).

38. "The Loft," Victorian Comprehensive Cancer Centre, http://www.theloftvccc.com.au/ (accessed April 28, 2017).

39. Nina Tandon interviewed for *TED Fellows Friday*, September 30, 2011, https://blog.ted.com/fellows-friday-with-nina-tandon/.(accessed February 2, 2017).

40. Ibid.

41. Clive Cookson, "Lunch with FT: Craig Venter," *Financial Times*, November 24, 2007, https://www.ft.com/content/f1cf9ccc-9700-11dc-b2da-0000779fd2ac(accessed April 10, 2017).

42. Fifty-nine percent of vegetarians are female and New York ranks fourth in the United States for the number of vegetarian restaurants after Portland, Seattle, and San Francisco. "Vegetarian Statistics," *Statistic Brain*, https://www.statisticbrain.com/vegetarian-statistics/ (accessed April 16, 2017).

43. Ashlee Vance, "Craig Venter on the Hail Mary Genome and Synthetic Meat," *Bloomberg*, August 10, 2013, https://www.bloomberg.com/news/articles/2013-08-08/craig-venter-on-the-hail-mary-genome-and-synthetic-meat (accessed April 17, 2017).

44. J. Craig Venter, "The Biological-Digital Converter, or, Biology at the Speed of Light," The Edge Dinner, Turin, July 10, 2012, https://www.edge.org/conversation/j_craig_venter-j-craig -venter-the-biological-digital-converter-or-biology-at-the-speed (accessed April 17, 2017).

45. *Wall Street* (film, 1987), directed by Oliver Stone, written by Oliver Stone and Stanley Weiser, starring Michael Douglas as Gordon Gecko.

46. Alan McKinlay and Philip Taylor, "Power, Surveillance and Resistance: Inside the 'Factory of the Future,'" in *The New Workplace and Trade Unionism*, ed. P. Ackers, C. Smith, and P. Smith (London: Routledge, 1996), 285.

47. Carol Milano, "Lab Management: The Human Elements," *Science*, March 12, 2010, http:// www.sciencemag.org/career_magazine/previous_issues/articles/2010_03_12/science.opms. r1000086 (accessed April 13, 2017).

48. Mitzi Baker, "A Year after Opening of Clark Center, Collaborative Projects (and Peet's Coffee) Brewing," *Stanford News Service*, November 1, 2004, https://news.stanford.edu/pr/2004/clark -1103.html (accessed May 4, 2016).

49. Robin Wander, "Stanford's Clark Center Celebrates First Decade," *Stanford Report*, October 3, 2013. https://news.stanford.edu/news/2013/october/clark-biox-decade-100413.html (accessed May 4, 2016).

50. R. Anthony Hyman, "Communicating Biology: The Design of a Molecular Cell-Biology Labo- ratory," in *Gentle Bridges: Architecture, Art and Science, Max Planck Institute of Molecular Cell Biology and Genetics*, ed. R. Anthony Hyman, Gerhard Mack, and Juhani Pallasmaa (Basel: Birkhäuser, 2003), 72.

51. Wieland B. Huttner, "Birth of a New Institute—Biopolis Dresden," *Nature Reviews Molecular Cell Biology* 2 (2001): 699–703, http://www.nature.com/nrm/journal/v2/n9/full/nrm0901_699a .html. (accessed July 1, 2016).

52. Tatjana Thelen, "Lunch in an East German Enterprise—Differences in Eating Habits as Sym- bols of Collective Identities," *Zeitschrift fur Ethnologie* 131 (2006): 55.

53. Ibid., 54.

54. Thorp and Goldstein, *Engines of Innovation*, 79.

55. A. Sturdy, M. Schwarz, and A. Spicer, "Guess Who's Coming to Dinner? Structures and Uses of Liminality in Strategic Management Consultancy," *Human Relations* 59, no. 7 (2006): 929–960.

56. Jeffery Sobal and Mary K. Nelson, "Commensal Eating Patterns: A Community Study," *Appe- tite* 41 (2003): 186.

57. "Joe Caters to You," joe coffee, http://joenewyork.com/catering/ (accessed July 3, 2016).

58. "Our Story," Riverpark, https://www.riverparknyc.com/our-story/ (accessed July 3, 2016).

59. Sam Sifton, "Intrigue by the East River," *New York Times*, December 7, 2010, http://www
.nytimes.com/2010/12/08/dining/reviews/08rest.html (accessed July 4, 2016).

60. "Our Story," Riverpark.

61. BioMed Realty: i3 La Jolla (Studio 216, Seattle, 2014), digital, https://vimeo.com/99157683
(accessed August 22, 2015).

62. Hyman, "Communicating Biology," 66.

63. Analiese Richard and Daromir Rudnyckyj, "Economies of Affect," *Journal of the Royal Anthropological Institute* 15 (1) (2009): 57–77.

64. Ibid., 57–77.

65. Chrispy, comments on Derek Lowe, "Fancy Building, Fancy Science," *In the Pipeline*,
AAAS *Science*, February 4, 2009, http://blogs.sciencemag.org/pipeline/archives/2009/02/04/fancy
_building_fancy_science (accessed August 19, 2016).

7   Naked in the Laboratory

Chris L. Smith

Jacques Derrida's posthumously published text, *The Animal That Therefore I Am* (2008), commences with an account of the philosopher emerging naked from a shower and being seen by a domesticated nonhuman animal. It's an awkward moment for Derrida. He describes the scene where he is "stark naked [*à poil*] before a cat."[1] He considers his own awkwardness and shame and the cat's nonchalance. For Derrida "the property unique to animals, what in the last instance distinguishes them from man, is their being naked without knowing it."[2] The sentences that follow suggest much of the logic of the text. Derrida starts his story with his cat but it comes to be an essay concerned for all nonhuman animals—but not all animals summed up and generalized under the term "the animal." Derrida is concerned for the particularity and specificity of nonhuman animals, for the different and diverse populations that are thrown together under the collective noun "animal." It is a concern that contemporary bioscience laboratories deal with every day. The scientists who, likely, leave their own cats and dogs at home work in spaces replete with other animals: rats, mice, guinea pigs, rabbits, zebrafish, fruit flies, and sometimes pigs, dogs, or monkeys. These animals are not, however, a concern that the architecture of laboratory buildings expresses clearly or openly. This concern generally comes to occupy basements and is always well concealed.

In this chapter I turn to a formally exuberant and scenically located biomedical research laboratory, the Parc de Recerca Biomèdica de Barcelona (PRBB, Barcelona Biomedical Research Park), designed by the Spanish architects Manuel Brullet, Albert de Pineda, and Alfonso de Luna. The building occupies a prominent location on a picturesque part of the coast of Barcelona. To the northeast is the Puerto Olímpico and a shimmering golden fish sculpture designed by Frank Gehry for the Barcelona Summer Olympics of 1992. To the northwest is the huge Parc de la Ciutadella in which sits the city's zoo, the Parc Zoològic de Barcelona, established in the late nineteenth century (and once famous for housing an albino gorilla). To the southwest is the adjacent Hospital del Mar, and to the southeast is a popular beach of golden sand, the Passeig

Figure 7.1
The PRBB viewed from the beach in 2013. Photograph by Sandra Kaji-O'Grady.

Marìtim. This part of the city was redeveloped as a port and casino, and to host the Olympic sailing races, and throughout summer, the beach is strewn with beautiful, bronzed, near-naked bodies.

The laboratory building, the PRBB, was constructed in 2006 as part of the post-Olympics consolidation of the waterfront. It is emblematic, in two senses of the word. First, the building is iconic: with its distinct elliptical shape it is a highly recognizable piece of architecture on a magnificent site. Second, the PRBB might serve as representative of any number of biomedical research facilities of the twenty-first century. The economies into which it is plugged, its rhetorical strategies, its architectural organizations and expressions, are shared with any number of laboratory buildings. It also engages the animal, the human animal, and the nonhuman animal, domesticated and undomesticated, in a manner similar to almost all biomedical laboratories. The difference in the PRBB that this chapter explores is that the juxtaposition between the human and nonhuman animal is peculiarly pronounced. The promoted lifestyle that the PRBB affords the scientists is in stark contrast to the life management to which the nonhuman occupants are subjected. In her book titled *Architecture, Animal, Human: The Asymmetrical Condition* (2006), Catherine Ingraham suggests that "architecture's primary response to biological life has been to organize its animate potential, to set it in motion down a path or through a program of occupation—to send it on its way through the world of the city, the site, the building."[3] As is the case in many biomedical research buildings, the nonhuman, standardized laboratory animal of the PRBB occupies a basement. It is unusual, however, that the laboratory animals share the basement with numerous lifestyle, sporting, and recreational facilities. This basement level continues under the road and plaza to connect with the adjacent beach and neighboring hospital. However, the lives that flow through the PRBB's basement do so in very different directions. Biomedical science similarly organizes the animal and sends it on its way, and the architecture of biomedical laboratories makes these journeys near-seamless.

Nude

Emerging from the shower, Derrida's encounter with his cat causes him to speak "from the heart";[4] an odd position for the philosopher most readily associated with deconstruction, the productive ambiguities, and political insecurities of language. The author of works such as *On Grammatology* ([1967] 1978) and *Writing and Difference* ([1967] 1978), Derrida knows well the difficulties of speaking from the heart. He diagnoses this problem within *Of Grammatology* as "logocentrism."[5] The word derives from the Greek *logos*, which translates as the direct relation between the "word" and the "world" (or "things"). Logocentrism is the belief in an ultimate reality or truth behind thought and thus relates to a desire for language to encompass truths; that is, for *words* to engender or accurately represent *things*. The endgame here is that words would *be* truth. Logocentrism would allow us to assert facts, to speak of things with absolute confidence. Logocentrism across the history of philosophy operates much like the scientific method across the history of science, assuring validity and efficacy to assertions. Derrida, however, articulates the despondency, indeed the malignancy, of this rationalist "nightmare." He engages the cruelties of applied rationality to expose the arrogance, or what Michel Foucault independently identifies as the "fascism," implicit in claims of reason.[6] Derrida's fixation with nonhuman animals also derives from the malignancy of logocentrism. To speak is considered a particularly human thing. To assert truth is an objective that places the truth-makers in a position of authority. There is something tragically anthropocentric about logocentrism. For Derrida, the self-confirming authority of the human prefigures the condemnation of the nonhuman animal.[7] Against the certainties of logocentrism and the political problematics of anthropocentrism, Derrida mobilizes a radical argument in noting that structures of meaning include and implicate any observer of them: to observe is to interact. Derrida's cat, standing *before* him, is an observer. Readers of science might note that this argument has a close correlate that is equally feline; the story of Schrödinger's cat.[8]

While the Austrian physicist Erwin Schrödinger considered the cat as a *subject* of scientific experimentation, both "alive and dead," for Derrida the nonhuman animal is itself the observer, but not an observer as the human might observe. For Derrida, the nonhuman animal is of an "absolute alterity"[9] and has "the vantage of this being-there-before-me."[10] The idea that the animal precedes (or indeed that the human follows) is one that Derrida plays out in much detail in *The Animal That Therefore I Am*. The animal comes "before" in the sense that "the animal" as a term is a mass summation of all animals. Thus, what is termed "the animal" has always been here. The animal comes before me, in a temporal sense. For Derrida, the animal also comes "before" in the sense that the animal has a very particular vantage point. The animal *sees* the human from a privileged point of view. Though Derrida doesn't use the word, what is suggested is a type of *nonchalance* to the gaze of the animal that falls upon the human.

When Derrida famously stated "there is nothing outside of the text" he implicated text in the semiological sense of extended discourse, that is, all practices of interpretation and communication that include, but are not limited to, language.[11] For Derrida any identity or meaning is provisional and relative, because it is never exhaustive, and it can always be traced back to a prior network of differences, and further back again.[12] When Derrida writes that he wishes to speak "from the heart" it is perhaps to this *prior network* to which he turns and for Derrida speaking from the heart, thus, is not to speak absolute truth but relative and contingent truths. Some forty years after *Of Grammatology* and *Writing and Difference*, the mode of writing that the philosopher engages in *The Animal That Therefore I Am* is in resonance with the topic to which he turns.

There is something raw and disarming about speaking from the heart, and Derrida speaks of nakedness in the same way. The philosopher asks of his cat what it is to be "naked without knowing it."[13] The question is really about the contingencies of human logics. For Derrida, the animal sits beyond the *logosphere* that concerns the human, the rational human animal, fixated on text, words, and communication. He is keenly aware of the issue of personification involved in approaching any animal, in ascribing any particular humanity or sensitivity to the animal. Derrida writes, "The animal, therefore, is not naked because it is naked. It doesn't feel its own nudity."[14] The idea of nakedness, Derrida concludes, is mounted only by those that clothe themselves: "From that point on … animals would not be, in truth, naked."[15] Derrida configures *knowing* with what it is *to be*, that is, with being. In this way epistemology is made relative to/ with/follows ontology. Derrida notes that "the animal would be *in* non-nudity because it is nude, and man *in* nudity to the extent that he is no longer nude. There we encounter a difference or *contretemps* between two *nudities without nudity*."[16] The nonhuman animal cannot be naked, or know of nakedness, or the odd shame and embarrassment associated with nudity.

Derrida, standing before his cat, his domestic animal, notes that the cat does not seem particularly discomforted by the interaction. He, himself, however has a sense of his own nakedness. It is a sense that comes with shame and some odd embarrassment that leaves him reaching for a towel (cloth) or his clothes. The shame that is felt in nudity is for Derrida coextensive with the development of clothing that hides or covers our sex. In this sense, only the human can be naked but the shame of nakedness is not an entirely interiorized shame. It is a shame that equally implicates cloth. For Derrida "clothing derives from technics. We would therefore have to think shame and technicity together, as the same 'subject.'"[17] This story of shame and technicity has a long history. Sigmund Freud associates the human species' walking upright with the exposure of genitalia and the development of a sense of shame. He proposes the shame associated with the exposure led to clothing. Derrida bypasses Freud's evolutionary account and reminds us of the mythical story of Prometheus. This tale of human and nonhuman attributes is also the story of the connection between nakedness and technologies:

Prometheus steals fire, that is to say, the arts and technics, in order to make up for the forgetfulness or tardiness of Epimetheus, who had perfectly equipped all breeds of animals but left "man naked [gymnon]," without shoes, covering, or arms, it is paradoxically on the basis of a fault or failing in man that the latter will be made a subject who is the master of nature and of the animal. From within the pit of that lack, an eminent lack, a quite different lack from that he assigns to the animal, man installs or claims in a single stroke *his property* … and his *superiority* over what is called animal life. This latter superiority, infinite and par excellence, has as its property the fact of being at one and the same time *unconditional* and *sacrificial*.[18]

Derrida proposes that such a relegation of the nonhuman animal "hasn't stopped being verified all the way to our modernity"[19] and for Derrida the very term "animal" invokes a "veritable war of the species."[20] This war is not the "tooth and claw" war depicted by Tennyson, as a war of one creature against another, but rather a war of one species against all others. It starts, according to Derrida, with the very act of collectivizing all that lives (with the exception of Man) under a single word. "The animal is a word," Derrida writes.[21] It is a word invoked in order to generalize all the (other) species, all the immeasurable diversity, under a single manageable term. Derrida concerns himself with what the translator of the text calls the "fragility and porosity of the supposed frontiers of the 'proper' upon which we have presumed for so long to found the traditional opposition of 'man' and 'animal.'"[22] Derrida reacts against the formulation of a world whereby "the animal" is spoken in the singular—as if all animals could be grouped as one against the human. His point is not to give extra privilege to domestic animals or animals that live closer to humans. His point is that the human animal might consider the particularity of nonhuman animals and not ascribe human characteristics to them. Derrida calls into question what he calls "this auto-biography of man"[23] and what it is "to name in general but in the singular, *the animal.*"[24] Donna J. Haraway makes a similar point in suggesting that "multispecies flourishing requires a robust nonanthropomorphic sensibility that is accountable to irreducible differences."[25] To differentiate between domestic and nondomestic animals is to merely create another category that is operative only in human terms. The *domos* of domesticity becomes as problematic as the *logos* of logocentrism. It seems to be a particularly poignant act of cruelty to invoke architecture and words against that which does not occupy the *logosphere.* In response, Derrida invents the term "animot." Phonetically indistinguishable from *animaux,* the plural of animal in French, "animot" speaks of the immense diversity of animals. But Derrida also invokes the term to draw our attention to the idea that "animal" is only a word (*mot*). He came to dismiss his own word "animot" almost as quickly as he invoked it. His concern in *The Animal That Therefore I Am* is not for words, but for the frontiers of the human and nonhuman animal, for the fragile boundary conditions and the porosity of this zone.

Near Naked

Haraway refers to "the entangled labor of humans and animals together in science."[26] Just as the philosopher might emerge from a shower naked before their cat, so too a scientist. Scientists retire to domestic spheres with cats and dogs and guinea pigs and rabbits. Their places of work are also full of animals. The laboratories of bioscience are a highly poised frontier of opposition and a knot of entanglement. They are a fragile boundary condition and a particularly porous zone—a highpoint of both "shame and technicity together." The humanity of the human is called into question in such zones, something that the PRBB unwittingly reveals.

The PRBB delights in the technical resolution of its own structural assembly, its material construction, its complex machinery of laboratory servicing and air-conditioning. It is a building whose technical quality and architectural aspirations were realized without being compromised by inadequate budget, banality of vision, or the timidity of city

Figure 7.2
The timber "veil" of the PRBB in 2013. Photograph by the author.

planners and regulators. The PRBB is the product of large-scale public investment and an ostentatious commitment to human health and well-being through research target-ing specific diseases with the new tools of genetic engineering. Like other biomedi-cal research laboratories in this book, it is a point of economic indulgence. Though, "indulgence" is perhaps not the right word. These buildings are framed by their own-ers, institutions, and residents as absolutely necessary. They are routinely described in bombastic terms, as investments for the future of humanity. International competition for the best scientists, public and private finance, and for the pharmaceutical dollar has led to many of these buildings becoming prominent in cities over the planet. The PRBB is no different in this regard. At a construction cost of USD 137 million (123 million euros), the PRBB was funded by the Catalan government, the City Council of Barcelona, Pompeu Fabra University, and the European Union. It opened in 2006 and currently houses seven independent research groups, with a recurring aggregate annual research expenditure of around USD 61 million (80 million euros).[27] The mis-sion statement of the PRBB is clear: "The Consortium PRBB contributes to create the best conditions so that the scientific community of the park is successful,"[28] where "success" tends to be measured in broad terms related to human health and the impact of research. The PRBB houses research groups working on a diverse range of projects "encompassing everything from organogenesis to finding treatments for neurodegen-erative diseases."[29]

With its courtyard facing the water, the PRBB relates more to the sea than it does to the surrounding buildings, but it is nevertheless conceived as a technical object. The architects use the descriptor "industrial object" to define the building and suggest it is much like the industrial objects of the nineteenth and early twentieth century as "single, unique, defined, impressive objects."[30] This depiction resonates with the defi-nitions of iconic objects more generally. Jordi Camí, the general director of the PRBB and a professor of pharmacology, refers to the PRBB as a "factory of the future."[31] The PRBB is a well-crafted and carefully articulated object that sits at the edge of an urban and urbane city. The term "factory" might indicate something of the productivity of the place, but it is no indication of the aesthetic of the architecture of the PRBB. The architects repeat words such as "fluidity, lightness and autonomy" in describing the PRBB.[32] The words are accurate descriptors. Producing a building that is 35,000 square meters (approximately 375,000 square feet) in area, and yet both "fluid" and "light" is a remarkable achievement.

The form of the PRBB is a truncated cone. It rises ten levels above the ground plane at its peak, and slopes down to six levels on the southeastern side facing the sea. Its plans are elliptical with a rectangular slot forming a courtyard that opens the building to the sky and the sea. The laboratories of the PRBB are arranged in an orthogonal and linear manner around the rectangular slot in something like a "U" shape. The U opens toward the sea and there are long corridors and open terraces running north–west to south–east. This orthogonal U-shaped arrangement of the laboratories yields repeatable and standardized laboratory configurations, and efficient horizontal and vertical circulation. The more flexible, open and social spaces and seminar, office, and write-up rooms and areas occupy the generous gaps between the angular U of laboratories and the fluid curve of the exterior ellipse. A notable point of difference between the PRBB and many other laboratories of this era is the amount of outdoor space and sunlight flowing into the building. Sunlight penetrates deeply and large decks and balconies are never far from the scientists even when in the laboratories because of the slot that cuts deep into the volume. The architects tell us that "the entire building revolves around this open space."[33] The PRBB owes Louis Kahn and the Salk Institute a debt for this spatial strategy. The architects of the PRBB had visited the Salk Institute in La Jolla, California, twice during the planning and design process.[34] Each of the upper levels of the PRBB contains a terrace that opens to the slot that is cut from the ellipse and thus opens to sea views. These terraces are like the decks or verandas of domestic structures; they are made of unpainted timber and furnished with deck-chairs and tables for outdoor eating. Outdoor staircases run between the terraces and the various levels, endowing the building with a relaxed and leisurely atmosphere.

The technical resolution of the PRBB is seductive. All spaces and details seem to have been carefully considered and, despite the scale of the building overall, most spaces here feel comfortable. Domestic. But not homelike, it's more like being in a grand yet carefully articulated holiday resort. Lest this suggests exclusivity, it needs to be pointed out that at ground level provision is made for passersby to take a diagonal shortcut from the metro station to the beach. The truncated cone that constitutes the volume of the PRBB doesn't reach the ground, but as the architects note "remains suspended by the shadow of a vast porch."[35] The screen of western red cedar timber slats that shrouds the building reaches down beyond the floor of the first level. The effect is to make the building appear to be less weighty, as if it were floating. The timber slats shade the structure and veil or cloak the functional parts of the building. The effect from the ground is pleasurably dizzying. The harshness of the sunlight is removed and yet one is compelled to look upward and enjoy the sky. It is a little like moving under the hem of a magnificent lacy skirt.

Shame and Nonchalance

The architects of the PRBB produced a sizable monograph on the building's development, design, and technical resolution. Camí wrote the introduction, in which he

Figure 7.3
The slot through the volume of the building to the ocean beyond. Photograph by UPF, Wikimedia Commons.

suggests two key functions, or values, that the architecture of the PRBB offers. The first is a sentiment that is repeated over and over again in the discourse of biomedical laboratory architecture. Camí writes, "above and beyond the technological aspect, the great purpose of this architecture is to provide spaces for relation and communication."[36] Architecture's role in the promotion of communication among the researchers is an oft-repeated refrain of architects, scientists, and the institutions involved in commissioning such projects. What is interesting, however, is that Camí goes on to qualify the idea and offers a second key function of the architecture. Camí suggests that the PRBB building "is an aesthetic attraction which lends dignity to scientific activity."[37] Such a raw concession is unusual in the discourse surrounding the biomedical laboratory typology. It begs the question: why must dignity be "lent" to biomedical science? It also begs the question: by what authority might architecture lend "dignity"?

Camí's suggestion resonates with the philosopher Georges Bataille's critique of architecture. In his short essay titled "Architecture" (1929), Bataille writes of "the architectural chain gang."[38] He is "against architecture" inasmuch as architecture operates structurally as a language, a geometry, and a mathematics that conceal the violence of

Figure 7.4
The elliptical screen in 2013. Photograph by the author.

authority and life behind an image of stability, order, and matter-of-factness.[39] Bataille famously suggests that architecture's key purpose is to "cloak" the violence of institutions. The dignity that architecture gives the PRBB may be such a cloak. Camí's introduction to the monograph on the PRBB prompts us to deal with Derrida's assertion that we "have to think shame and technicity together, as the same 'subject.'"[40] It is from this particular position, or in respect to this specific moment, that I wish to look a little deeper into the PRBB, with its scientists and exceptional technicity, and its mice, rats, rabbits, frogs, and zebrafish.

As of 2016, the scientific community of the PRBB included 1,468 "residents": senior researchers, postdocs, PhD students, research technicians, support service and administrative staff, and visiting researchers. The published data suggests that between 24 and 31 percent are from outside Spain; 43 percent are male, 57 percent are female.[41] No statistics are kept on whether or not the scientists keep dogs or cats at home, but if the scientists conform to the averages of pet ownership in Spain we can assume 26 percent of households have one dog or more and 19 percent one cat or more.[42] But this is an assumption. Information from the "animal facility" of the PRBB is less expressive and less exact than that provided on the "scientific community." The PRBB suggests that the animal facility "has capacity for 70,000 mice under SPF [specific-pathogen-free] conditions as well as 50,000 zebrafish."[43] There are also rats (of unknown number) and 400 frogs here.[44] It follows that nonhuman animals in the PRBB may outnumber the humans 80:1. Overall the human is a rarefied and rare commodity here. It is noteworthy that numbers relating to the nonhuman animals tend to be rounded. (Sometimes to what appears to be increments of 5,000). The PRBB annual report of 2016 notes: "Exports and imports of animals worldwide: 700"; "Mouse biopsies for genotyping: 25000"; "Blood tests carried out: 1000"; "Gamma radiation carried out on cells and mice: 800."[45] Other than mice, no species are identified in the annual report. The report also notes: "Participants in accreditation courses for researchers in the use of experimental animals: 35."[46] One assumes these thirty-five were human. This number seems extraordinarily low given the number of scientists operating at the PRBB.

The inexact character of the numbers relating to the animals of the PRBB is fascinating, given the PRBB operates an exacting animal-facility data management system for monitoring the animals. A report on the data management system suggests "each animal's profile will include for example age, strain, matings, and any litters the animal has borne, room where the animal is and the IACUC procedure related to each one."[47] (The human resources section of the PRBB could only dream of collecting such information.) Mireia Juan, a senior PRBB Animal Facility supervisor, reported that the monitoring system operated by the PRBB "help[s] us stay in control of what is happening to every animal."[48] While individual animals come to occupy spreadsheets and digitized management systems that allow a careful monitoring of every moment of their basement lives, they are identified only by codes. That is, the animals are not named. Derrida writes, "it seems to me that every case of naming involves announcing a death to come in the surviving of a ghost, the longevity of a name that survives whoever carries that name."[49] We name the animals closest to us—our cats and dogs—for which we care. The longevity of the nonhuman animal of the biomedical laboratory, however, was never part of the agenda.

Below the suspended elliptical floors of the upper nine levels of the PRBB, and well below the skirt of the building that floats seductively above the ground plane, lie another three levels. One is partially buried and two are completely underground. The lifestyles of the scientists of the PRBB are enhanced by the provision of sporting and recreational facilities in the basement levels. Occupying all three levels is a sports center, with two tiers of seating for 300 people, three basketball courts, and thalassotherapy pools. The incorporation of sporting and lifestyle facilities at the PRBB came with considerable expense and technical difficulty. Large open areas at basement level meant that the structural system of the PRBB had to operate with large spans (you couldn't have a column protruding through a basketball court) and unconventionally the entire above ground structure came to be suspended from the roof.[50] This substantial sporting area is linked to the beach by a tunnel that passes under the adjacent road. The scientist, changing out of a white coat and into garishly bright spandex swimming briefs, can jog from the sports center directly onto the beach opposite. People walk dogs along the promenade, and dogs are allowed on the beach outside of the summer months (from the first of July until the last Sunday in September). This beach is also a particular point of attraction for the scientists of the PRBB. The PRBB holds a Beach Volleyball Championship with thirty-two teams playing six-on-six games and twenty-two teams playing in teams of four. In total, over five hundred PRBB employees participate in the games that, from April to July, number three or four daily. The volleyball championship culminates in an annual summer party with awards ceremony, live music and performances, and sandwiches and beer.[51] The event is even professionally filmed and uploaded to YouTube.[52]

There is also a theater group of PRBB employees, as well as yoga enthusiasts, a choir, the Afro-Brazilian martial arts capoeira, and an African dance and percussion group. Again, these concerts and performances are posted on YouTube. Witnessing PRBB scientists singing, partying, and playing together, it is easy to forget the concerted efforts made by the corporation, and for the corporation, to cultivate collective recreational and leisure aspects of the scientists' lifestyles. The value of the collective over the individual is emphasized in the social and sporting drives of the PRBB. One is reminded of Isabelle Stengers's description of the scientist as something of a pack animal and science as complex "collective practices."[53]

The basement levels of the PRBB also contain loading bays, along with a large carpark and the electricity and water supply systems. On basement level –1 there is a synchotron and radiopharmacy with a direct below-ground connection to the Hospital del Mar, as well as the hospital's own animal facility. (It might have been equally simple to link the animal facility to the adjacent zoo, but this was never part of the agenda.) The PRBB's animal facility has been accredited by the Association for Assessment and Accreditation of Laboratory Animal Care AAALAC since 2010. The PRBB also operates an Ethical Committee for Animal Research (CEEA) and it must be noted that the PRBB is remarkably frank in the information it makes publicly available about its engagements with nonhuman animals.[54] Information is published as part of its annual reports and it is a simple matter to upload the PRBB's service and pricing lists related to its animal facility. There are prices listed for everything from "Tail cut, Ear tag" (1,16 euro/item) through to "Mice ovary transplantation" (125,83 euro) and "Embryo cryopreservation" (2,101,19 euro/line).

On the one hand, we might be grateful for the openness of the information. On the other, we might be left feeling some shame or discomfort at a disregard for the particularity of individual animals. Of course, the failure to sacrifice the animal comes with danger. The biblical story of God's rejection of Cain's offering of fruit, and his acceptance of Abel's sacrifice of a lamb, and the recurrence of human and animal sacrifice in Greek mythology, might attest to this risk. It follows that much has been said of the value of the animals sacrificed in the name of human health and well-being. Derrida does not enter into the discourse related to the question "is it worth it?" He does not enter a discussion on sacrifice, excess, and value systems (as Bataille had before him). And Derrida also steers well clear of questions such as "what does an animal feel?." This discourse always ends in a complex personification of the animal—equating its thoughts and perceptions with ours. It ends with us trying to humanize some animals, the domestic animals (and dehumanize those we are willing to sacrifice). This would be to make the mistake of assuming that animals are naked. Rather, Derrida posits a simple question that sits beyond the personification of the nonhuman animal. He asks simply: "Can they suffer?"[55] and suggests that in this question there is "no room for doubt."[56]

The animal facility of the PRBB occupies an area of approximately 4,500 square meters. There are six key areas: an "aquatic area" (which contains *xenopus* frogs and zebrafish); what is known as a "barrier zone" of pathogen-free (SPF) space with three support laboratories; a "transgenesis area" composed of two large laboratories (each of approximately 250 square meters) "to apply both traditional and the most innovative techniques, able to generate genetically modified mice";[57] and an "experimental area" that is "equipped with laboratories for *in vivo* imaging, rooms for behavioral studies, surgery rooms and an irradiator."[58] An irradiator is not for eradication (well, not entirely); it is for the application of gamma radiation. The facility also has an area referred to as the "quarantine area" "with its own locker room access, rederivation laboratory for embryo transfer, laundry and sterilization, and four different rooms with modern microisolators," and what is referred to as the "conventional animal house." In the literature outlining the various services that the PRBB provides its in-house clients (and retails to external organizations) the conventional animal house space is described as being "located in the adjacent building … and connected to the PRBB building via an underground tunnel. It has an area of 1,300 m2 for neuropharmacology and immunology studies."[59] While Derrida notes the uses to which "the animal" has been put (sacrifices, transport, ploughing, small-scale butchering, experiments on the animal), he turns his attention to the more recent change in the manner in which the animal is handled:

It is all too evident in the course of the last two centuries these traditional forms of treatment of the animal have been turned upside down by the joint developments of zoological, ethological, biological, and genetic forms of *knowledge*, which remain inseparable from *techniques* of intervention *into* their object, from the transformation of the actual object and from the milieu and world of their object, namely, the living animal.[60]

In order that the PRBB might "stay in control of what is happening to every animal," the animal facility is highly secure and painstakingly removed from any outside interference.[61] Sunlight, natural ventilation, the ground, the sky, and the sea are, in such circumstances, potential sources of interference. The services associated with the animal facility are highly particularized. Temperature, humidity, ventilation, and lighting are strictly controlled. Lighting is controlled to "simulate the cycles of the sun."[62] The lightness and fluidity of that portion of the PRBB that sits above ground and that relates to the human animal is oddly *simulated* in the nonhuman animal facility; the PRBB does not just *house* the animal. "Every step with an animal—surgery, anaesthesia, and necropsy, among the dozens of others—takes place within the animal facility's lab."[63] The animal facility is also instrumental in creating animals. The transgenesis area composed of large PC2 (physical containment level 2) laboratories is, perhaps, the most tightly controlled space within the facility. This space is described in different ways in different forums. The publicly accessible PRBB online services directory, which serves to indicate what services may be provided (purchased) from the PRBB, suggests the area is "to apply both traditional and the most innovative techniques, able to generate genetically modified mice."[64] An account of the consultant engineer Joan Gallostra from Group JG that worked on the building and helped remove the contingencies of temperature variation, airflow, and sunlight from the animal facility, is equally brief about what occurs in the transgenesis area. While the technical specification from the engineer was exacting and his skill in controlling basement environments is evidenced, his description of the use of the transgenesis area is vague. Gallostra suggests the transgenesis area is "an area for studying behavior with two equipped operating rooms."[65] Gallostra does not name the animals involved but is secure in the knowledge that they'll never see the sunlight or him. For Derrida, "the gaze called 'animal' offers to my sight the abyssal limit of the human: the inhuman or the a-human, the ends of man the bordercrossing from which vantage man dares to announce himself to himself."[66]

Naked

Derrida notes that for philosophy and philosophers, "the experience of the seeing animal, of the animal that looks at them, has not been taken into account in the philosophical or theoretical architecture of their discourse."[67] The point is all the more poignant for science and the scientist. The cats and dogs sharing their homes are regarded with very different eyes than those that focus upon the mice and rats of the laboratory. Schrödinger's cat might attest to this fact. It should be noted that the tensions that operate regarding the nonhuman animal and that are expressed by animal activists, philosophers, and scientists are complex, and it is not my intent to oversimplify the positions. The use of nonhuman animals in a laboratory is the product of training, protocols and socialization, investment, and a clear (sometimes noble) focus on human health and well-being; it is not a simple demonstration of human estrangement from the animal. The motto of the contemporary biomedical research facility is "from benchside to bedside." The slide of logics, economics, and value is from laboratories to clinics and hospital wards. But no experimental medical product or process proceeds without the mediating animals that live between experimental hypothesis and patient. The motto carefully bypasses the cages and the bodies of nonhuman animals that are routinely sacrificed in this process. The animal is dealt with as an object of the science—an extension of the bench. It is placed and positioned in the discourse of the contemporary laboratory very carefully as one might place a numeral in a spreadsheet. When the nonhuman animal enters discourse, it tends to do so as an act of personification (where personification refers to a making-human of the animal), but it is a particular personification; in laboratory parlance, the animal is spoken of as a "sacrifice" and the "sacrificed." The sacrifice of the nonhuman animal is not for itself of course, but for us. Derrida refers to this as "an immense disavowal"[68] and Haraway, reflecting on Derrida's work, describes it succinctly as "the unresolved dilemmas of killing and relationships of use."[69] Derrida is clear that the contemporary manipulations of the animal mark a shift in the means by which the animal comes to be activated/decimated/manipulated, what Derrida refers to as the "subjection" of the animal. This includes

by means of genetic experimentation, the industrialization of what can be called the production for consumption of animal meat, artificial insemination on a massive scale, and more and more audacious manipulations of the genome, the reduction of the animal not only to production and overactive reproduction (hormones, genetic crossbreeding, cloning, etc.) of meat for consumption, but also of all sorts of other end products, and all of that in service of a certain being and the putative human well-being of man.[70]

One wonders if this might be what Camí had in mind when he suggests that the scientific activity of the PRBB needed to be lent "dignity."[71] I remain unconvinced that architecture might grant this desired dignity. At best, the architecture of the PRBB successfully cloaks the nonhuman animals of scientific research. It successfully hides, carefully controls, and ingeniously removes these animals from sight, as they are removed from sunlight. The skirt of this building, the clothing in which it is covered, is a wondrous cloak that keeps visitors focused upward to the sky and outward to the sea. No one is looking too closely into the undercarriage for other animals. It is a secret in which we are all complicit. The entire biosciences industry rests on a rationale that distances us from and erases the animals whose lives we value less than our own. The lifestyles we enjoy—from the health and longevity of life that we have, to the cosmetics that make even our faces less naked—rest on a largely unspoken debt. The PRBB, like all bioscience laboratories, thrusts lifestyles and lives together. But there is nothing balanced about the arrangement. In a book of over two hundred pages, the architects of the PRBB dedicate this sentence to the animal facility: "Also on this underground floor are the lab animals [sic] quarters, a vast centre with the guinea pigs used in the research done in the Biomedical Research Park that is also linked underground to Hospital del Mar on the north side."[72] It is hard to tell whether the architects' reference to "guinea pigs" is to the broader idea of a guinea pig in the colloquial sense of any animal that is subject to experimentation, or whether they genuinely believed the animal facility to be populated by the species. Oddly, more words are dedicated in the book to the durability of the timber used on the cloak of the building than the entirety of the animal facility.[73] The architects and engineers remain nonchalant. Before his cat, Derrida reminds us that it is not principally a question of how we see the animal, it is a question of how the animal sees us: "The animal looks at us, and we are naked before it."[74]

Notes

1. Jacques Derrida, *The Animal That Therefore I Am* (New York: Fordham University Press, 2008), 4.

2. Ibid., 4–5.

3. Catherine Ingraham, *Architecture, Animal, Human: The Asymmetrical Condition* (London: Routledge, 2006), 20.

4. Derrida, *The Animal That Therefore I Am*, 1.

5. Jacques Derrida, "The End of the Book and the Beginning of Writing," in *Of Grammatology*, trans. Gayatri Chakravorty Spivak (Baltimore: Johns Hopkins University Press, [1967] 1976), 6–26.

6. Jacques Derrida, *Writing and Difference*, trans. Alan Bass (London: Routledge & Kegan Paul, [1967] 1978). Michel Foucault in "Introduction" to Gilles Deleuze and Félix Guattari, *Anti-Oedipus: Capitalism and Schizophrenia*, trans. Robert Hurley, Mark Seem, and Helen R. Lane (Minneapolis: University of Minnesota Press, 1983), xii–iv.

7. For an extensive exploration of Derrida's engagement with the nonhuman animal, refer to Leonard Lawlor, *This Is Not Sufficient: An Essay on Animality and Human Nature in Derrida* (New York: Columbia University Press, 2007).

8. Erwin Schrödinger, "The Present Situation in Quantum Mechanics: A Translation of Schrödinger's 'Cat Paradox' Paper," trans. John D. Trimmer, in *Proceedings of the American Philosophical Society* 124 (1980): 323–338.

9. Derrida, *The Animal That Therefore I Am*, 11.

10. Ibid.

11. Derrida, *Of Grammatology*, 158. Derrida's own explanation of the statement can be found in Jacques Derrida, *Positions*, trans. Alan Bass (Chicago: Chicago University Press, [1972] 1981), 65; and Jacques Derrida, *Dissemination*, trans. Barbara Johnson (Chicago: Chicago University Press, [1972] 1981), 261.

12. This turn may be toward something like what Roland Barthes describes as the "zero degree" of writing. Barthes calls the zero degree of writing a closure, a retreat, and a suspension of meaning. Roland Barthes, *Writing Degree Zero* (London: Jonathan Cape, [1953] 1967).

13. Derrida, *The Animal That Therefore I Am*, 5.

14. Ibid.

15. Ibid.

16. Ibid.; emphasis in original.

17. Ibid.

18. Ibid., 20; emphasis in original.

19. Ibid., 21.

20. Ibid., 31.

21. Ibid., 23.

22. Marie-Louise Mallet, "Foreword," in Derrida, *The Animal That Therefore I Am*, x–xi.

23. Derrida, *The Animal That Therefore I Am*, 24.

24. Ibid.

25. Donna J. Haraway, *When Species Meet* (Minneapolis: University of Minnesota Press, 2008), 90.

26. Ibid., 80.

27. Reimund Fickert, director of projects, PRBB, workshop presentation on the design of Campus Albertov, at Charles University, Prague, 2013, https://kampusalbertov.cuni.cz/KA-62-version1-prbb_kampus_albertov_conferenc.pdf (accessed March 28, 2017).

28. "PRBB Performance Report of 2016," https://prbbperformancereport2016.wordpress.com/home/ (accessed May 1, 2017).

29. Helen Kelly, "Designing and Managing the LIMS at PRBB's Animal Facility," *ALN*, May 28, 2015, https://www.alnmag.com/article/2015/05/designing-and-managing-lims-prbbs-animal-facility (accessed May 20, 2017).

30. Manuel Brullet, Albert de Pineda, and Alfonso de Luna, *Parc de Recerca Biomèdica de Barcelona (PRBB)* (Barcelona: Edicions de l'Eixample, 2007), 26.

31. Jordi Camí, "Welcome," Barcelona Biomedical Research Park, https://www.prbb.org/parc#benvingud.

32. Brullet, de Penida, and de Luna, *PRBB*, 23.

33. Ibid., 29.

34. Jordi Camí, in Brullet, de Pineda, and de Luna, *PRBB*, 12.

35. Brullet, de Pineda, and de Luna, *PRBB*, 23.

36. Camí, in Brullet, de Pina, and de Luna, *PRBB*, 12.

37. Ibid., 13.

38. Georges Bataille, "Architecture," in *Encyclopædia Acephalica: Comprising the Critical Dictionary and Related Texts*, Atlas Archive, 3 (London: Atlas, 1995).

39. Chris L. Smith, *Bare Architecture: A Schizoanalysis* (London: Bloomsbury, 2017), 61.

40. Derrida, *The Animal That Therefore I Am*, 5.

41. "Performance Report of the Barcelona Biomedical Research Park Consortium," *Barcelona Biomedical Research Park*, March 27, 2017, https://prbbperformancereport2016.wordpress.com/2017/03/27/data-and-facts/ (accessed May 11, 2017).

42. "Share of households owning at least one cat or dog in Spain from 2010 to 2014," *Statista*, https://www.statista.com/statistics/517042/households-owning-cats-dogs-europe-spain/ (accessed May 10, 2017).

43. "Animal Facility (AAALAC Accredited Unit)," Barcelona Biomedical Research Park, http://www.prbb.org/serveis. (accessed May 11, 2017).

44. Kelly, "Designing and Managing."

45. "Performance Report of the Barcelona Biomedical Research Park Consortium."

46. Ibid.

47. Kelly, "Designing and Managing."

48. Ibid.

49. Derrida, *The Animal That Therefore I Am*, 20.

50. Brullet, de Pineda, and de Luna, *PRBB*, 24.

51. Fickert, workshop presentation on the design of Campus Albertov.

52. Minifilms, *PRBB-BeachVolleyball Championship 2011 and Summer Party*, directed by Ramon Balague, https://www.youtube.com/watch?v=uicD-2NM8hM (accessed January 18, 2017). (For 2010, see https://www.youtube.com/watch?v=fZyuEy7ARM4.)

53. Isabelle Stengers, *The Invention of Modern Science,* trans. Daniel W. Smith (Minneapolis: The University of Minnesota Press, 2000), 106 and 134.

54. This level of openness is also evidenced in the PRBB participation in a 2017 television report on "Animals under the Microscope," for A la carta Televisión y Radio, http://www.rtve.es/alacarta/videos/repor/repor-animales-bajo-lupa/3931289/ (accessed January 21, 2017).

55. Derrida, *The Animal That Therefore I Am*, 28.

56. Ibid.

57. "Animal Facility (AAALAC Accredited Unit)."

58. Ibid.

59. Ibid.

60. Derrida, *The Animal That Therefore I Am*, 25; emphasis in original.

61. Kelly, "Designing and Managing."

62. Joan Gallostra, in Brullet, de Pina, and de Luna, *PRBB*, 136.

63. Kelly, "Designing and Managing."

64. "Animal Facility (AAALAC Accredited Unit)."

65. Gallostra, in Brullet, de Pina, and de Luna, *PRBB*, 136.

66. Derrida, *The Animal That Therefore I Am*, 12.

67. Ibid., 14.

68. Ibid.

69. Haraway, *When Species Meet*, 79.

70. Derrida, *The Animal That Therefore I Am*, 27.

71. Camí, in Brullet, de Pina, and de Luna, *PRBB*, 13.

72. De Luna, in Brullet, de Pina, and de Luna, *PRBB*, 38.

73. Francesc Xurigué, in Brullet, de Pina, and de Luna, *PRBB*, 117.

74. Derrida, *The Animal That Therefore I Am*, 29.

8   Biosphere 2's Experimental Life

Tim Ivison, Julia Tcharfas, and Simon Sadler

Biosphere 2, a 3.14-acre glass and steel enclosure built in 1991 in the middle of the Arizona desert, was originally designed as an experimental laboratory in which eight humans were to live within a closed-system environment of seven distinct ecological biomes, studying the changes that occurred within those various biomes, maintaining the system's infrastructure, and monitoring their own physiology and experience. The experiment was conceived on the dual basis that it would test the feasibility of long-term human settlement beyond the Earth's biosphere (Biosphere 1) and that it would model in real time the complex ecological and atmospheric adaptations between humans and a complex ecosystem. Now, more that twenty-five years after the first air-lock was closed, Biosphere 2 remains a unique experiment in complex systems, which has yet to be repeated at the same scale or complexity.

Biosphere 2 has recently resurfaced in the contemporary environmental imagination as a kind of cultural and technological artifact. The project appears as a challenge to the institutional conventions of ecological thought and a premonition of our current climate predicament. The emergence of anthropogenic climate change as a key crisis of the early twenty-first century, as well as a concomitant geological (and indeed geopolitical) discourse on the Anthropocene have forced a widespread reevaluation of human-ecological relations. Biosphere 2 remains one of the most ambitious attempts to model these relations in an experimental setting, prefiguring what is now largely a global computational practice. Historians of science have returned to Biosphere 2 as a case study of the fringes of ecological thought at the turn of the millennium,[1] while artists have begun to revisit the project site as a late manifestation of the counterculture and as a prefiguration of network culture—what Fred Turner has called the move "from counterculture to cyberculture.'"[2] Meanwhile, space science has resurfaced in the mainstream cultural discourse, with a number of high-profile closed-systems projects in recent years, from the Mars-500 collaboration between Russia, ESA, and China, to NASA's Hi-SEAS mission in Hawaii, not to mention private ventures such as SpaceX,

Figure 8.1
Biosphere 2 in Tucson, Arizona, built by Space Biospheres Ventures. Photograph by Gill C. Kenny,
ca. 1991.

Virgin Galactic, and MarsOne; each promises space travel and colonization missions in
the near (but often postponed) future.

The project has also reemerged in critical terms, characterized (or caricatured) as
mechanistic and anthropocentric in its approach to environmentalism,[3] or as a cautionary tale of overzealous corporate techno-science (or both).[4] Those familiar with the
mass media account of Biosphere 2 will remember that the project was at one point
entirely overshadowed by a mythology of tabloid intrigue, and accusations of misman-
agement and pseudoscience.[5] Throughout the long debate over problems in Biosphere
2, one particularly consistent line of criticism has been directed at what we might call
the folly of its humans—of its "manned mission." The alleged failure of Biosphere 2
has been ascribed to the psychodrama of putting eight people in an airlocked building
for two years, whose presence (and the priority put on their well-being) continually
undermined more "serious" or profitable ecological research.[6] Most troubling to the
mainstream press was the fact that Biosphere 2 was largely run by self-taught ecologists,
members of an organization called the Institute of Ecotechnics. Journalists reported that
this collective of environmentalists and experimental theater practitioners apparently

had the audacity (and, coincidentally, the resources) to circumvent the normal barriers of entry to institutional scientific research.[7]

This chapter instead attempts to bring the manned mission within Biosphere 2 back to the center of analysis, and to recover its radical implications for ecological research and thinking. We will argue that the most problematic aspect of the Biosphere 2 experiment—the human experiment—has also turned out to be its most important contribution to our ecological imaginary. The chapter argues that the manned mission attempted a new kind of laboratory praxis (or lifestyle, if you will) through which the participants came to understand, in a performative sense, an open-ended entanglement and interdependency with natural environments in a closed system. Ultimately, the controversy around Biosphere 2 can be linked directly to this experimental and durational quality of the manned mission and its divergence from the commercial and institutional norms of climate research. The premature curtailment of the manned missions and the abrupt change in management that occurred at Biosphere 2 demonstrate all the more clearly the unique model of scientific practice proposed in the first three years of the project.

Building the Biosphere

The idea for Biosphere 2 was originally conceived by the Institute of Ecotechnics (IoE), a transdisciplinary research group formed in 1973 to study human impact on global ecosystems. Cofounders John Allen, Mark Nelson, and Tango Parrish described the nascent discipline of ecotechnics as "extending the subject matter of ecology past its usual limits of studying floral and faunal population, their environmental resources (mineral and gas cycles), and their interplay, to include human, mechanical, chemical, cultural, and decision-making populations into the 'equation.'"[8] IoE held a series of annual conferences on ecological sciences beginning in 1974, each one addressing a different concern of ecotechnic research. These conferences attracted a diverse range of figures from across culture and the sciences, including Director of the Royal Botanic Kew Gardens Dr. Ghillean T. Prance, experimental ecologist Clair Edwin Folsome, Gaïa theorists James Lovelock and Lynn Margulis, marine biologists Hank Truby and Dr. John C. Lilly, Richard Dawkins, William S. Burroughs, R. Buckminster Fuller, mathematician Ralph Abraham, astronauts Joe Allen and Rusty Schweickart, and a host of artists, jazz musicians, and ethnobiologists.

The first five annual conferences considered distinct Earth biomes: deserts, oceans, grasslands, rainforests, and high mountain ranges; followed by five conferences addressing the Earth, the solar system, the galaxy, and the cosmos, and culminating in

Figure 8.2
Mission 1 crew: (top row from left) Taber MacCallum, Sally Silverstone, Linda Leigh, Mark van Thillo, and Mark Nelson; (bottom row from left) Jane Poynter, Abigail Alling, and Roy Walford. Photograph by the Institute of Ecotechnics, 1991.

a conference on the Biosphere in 1984. It was at The Galactic Conference in 1982 that the architect and IoE member Phil Hawes proposed that a model space base might put to work many of the concepts being explored by the group. The 1984 conference on the Biosphere was not a conceptual exercise, but an action plan, addressing this question: Was it materially possible to construct a self-regulating living system sealed inside a structure to house a crew for a century? By 1987, the conference on closed ecological systems brought together a team of scientific specialists who would help conceptualize, design, and consult on the project, including the Institute of Biomedical Problems in Moscow, ecologists Eugene and Howard Odum, and William Chaloner, head of the Geosphere-Biosphere committee of Great Britain. To fund the project, members of the IoE created a new partnership between one of the co-directors, Edward P. Bass—an investor and industrialist represented by his company Decisions Investment—and other members of the institute, resulting in the creation of Space Biospheres Ventures (SBV).

Informing the institute's early project research was a series of postwar biomedical experiments conducted by both Soviet and American space agencies to test the viability of manned space flight. In Russia, the BIOS-3 program was designed for three people to occupy a hermetically sealed cabin for up to 180 days, surviving mostly on an oxygen and carbon dioxide respiratory exchange between the human inhabitants, chlorella blooms grown in trays, and greenhouse crops of wheat and vegetables. Limited experiments had also been carried out by NASA, which consulted with Eugene and Harold Odum on the possibility of including complex ecological environments within space cabins as life support systems.[9] The IoE was also interested in Clair Edwin Folsome's invention of the ecosphere in the 1960s, a completely sealed microhabitat for shrimp, bacteria, and algae, which he discovered in the course of oceanic studies. Using data from Folsome's ecospheres, as well as insights drawn from BIOS-3, the Odum's research for NASA, and intensive farming data, the IoE engineer Bill Dempster and ecologist Mark Nelson calculated that three acres was the area necessary to produce enough food and oxygen as well as recycle waste and water for eight humans over an extended period.[10] Hence the closed system itself would be scaled to the number of its inhabitants.

The Biosphere 2 complex, designed by architects Margaret Augustine, Phil Hawes, and Peter Pearce, employed a unique steel-and-glass truss system adapted from Buckminster Fuller's geodesic domes to create a durable, airtight enclosure. Within the three-acre area, seven unique ecological "biomes" were designed for inclusion: a five-story human habitat (or "city" biome), an intensive agriculture area, a mountain rainforest with waterfall, a savanna grassland, an ocean with coral reef, a freshwater marsh, and a fog desert. These complex biomes, populated with thousands of insects, mammals, fish, birds, and plant species would be regulated by a coordinated "technosphere," including the passive solar energy of the Biosphere spaceframe, a system of overhead sprinklers and misting machines, a subterranean array of hydraulic pumps, water reservoirs, tunnels, and fans, as well as external cooling towers, variable expansion chambers (or "lungs") and a 5.2-megawatt power plant. An eight-person crew would be selected and trained to live within and maintain the closed system for two-year periods across the next 100 years.

In the abstract, this vast ensemble of technical and biological components can be understood in a number of ways. On the one hand, we might imagine it as a scaled-up version of Folsome's ecosphere, in which the materially enclosed, energetically open system of the Biosphere creates its own alien ecosystem. On the other hand, one might view the variety of different Earth systems brought into a single structure as a scaled-down representation of our own biosphere. The Biosphere 2 complex invited both of these functional and representational interpretations, oscillating between an open-ended, hybrid, performative exercise in complexity, and a totalizing symbolic system of relations, hierarchies, and cultural values all under one roof. It is tempting to see this closed system at the center of the Biosphere 2 experiment as a holist, harmonious image of environmentalism, but closer analysis reveals a much stranger scenario that asks more questions than it answers. Indeed, we might understand ecotechnics and the Biosphere 2 project in similar terms to those ascribed to Stewart Brand's notions of "whole design" and "co-evolution": "a quest for an ecological metaphysics that asked questions about our being in the world."[11]

In a more cautionary reading of these concepts, the historian of science Peder Anker has argued that Biosphere 2 could be seen as the "culmination of a tradition of research into ecological colonization of both outer and earthly space," a tradition he articulates with the concept of "cabin ecology."[12] According to Anker, cabin ecology emerged out of 1950s astronautics research, where it described the use of ecological principles to incorporate earthlike environments into the design of space vehicles, enclosures, and capsules. Howard and Eugene Odum, who would later be participants in the IoE conferences, articulated the two basic principles of these ecology-based life support systems. The first was "carrying capacity": a calculation of the interior space, resources, and breathable atmosphere needed to support a certain number of astronauts.[13] The other was that this cabin ecology would require some "little piece of this biosphere" that would be "materially closed but not closed to energy flux."[14] Anker notes that the Odum brothers developed cabin ecology applications for NASA in the 1950s and 60s, and furthermore, that they went on to serve as consultants on Biosphere 2, along with other "cabin ecologists" such as Buckminster Fuller, Lynn Margulis, and Dorian Sagan. For Anker, this link between the discipline of ecology and the military-industrial complex, demonstrated in the history of the idea and etymology of the term, led to an overly functionalist and anthropocentric theory of the environment, limiting (literally boxing) our conceptual discourse and creating a "managerial culture of scientific technocracy among environmentalists."[15]

Although the notion of cabin ecology serves as a useful periodization of Cold War and postcolonial ecological attitudes, we would argue that Anker's argument might be too narrow to account for the plenitude of Biosphere 2. As Anker notes, the cabin ecology principles developed and promoted by ecologists such as the Odums were ultimately jettisoned by NASA, deemed unnecessary for the short-term missions and goal-oriented budgets of the space program. The ecological approach of integrating biological and technical systems into life support far exceeded the requirements of NASA and the military. Instead, they relied on synthetic rather than biological systems to maintain life support during space flight.[16] Indeed, the internal composition of Biosphere 2 lacked both the strategic remit of a Cold War/Space Race project and the organizational logic required for such a venture. And if Biosphere 2 was the culmination of a cabin ecology more narrowly defined, then it was a strange cabin indeed. In terms of its internal composition, Biosphere 2 included seven distinct biomes—an excessive scenario compared to precedents of cabin ecology. Official publications point to didactic reasons of representation or demonstration, and to the idea of producing "a rich atmosphere" for the human inhabitants, but the expense of these biome habitats—and the sheer scale of Biosphere 2's ensemble—make any purely functional rationalization difficult to sustain. As scientific journalist Rebecca Reider writes, "Why include a fragile coral reef in the ocean, a waterfall gushing into a hidden pool in the rainforest, the sacred Amazonian hallucinogenic vine *ayahuasca*, or perfectly timed waves lapping against a sandy beach in the ocean? Practical considerations only told part of the story."[17]

In fact, there were two schools of thought about how to design and construct the wilderness biomes inside Biosphere 2. On the one hand, very particular species of plants and animals were being selected for inclusion in each of the wild biomes. On the other hand, elements such as the seawater from San Diego, the coral reef from the Caribbean, and "marsh modules" containing mangroves, mud, and microbes from the Florida Everglades were also entered into the system. Curt Suplee summarized in the *Washington Post* how "One faction favored massive redundancy, packing in enough species so that some were bound to survive; the other urged using restricted numbers of species preplanned for optimal coexistence. Eventually, both groups coalesced with the result that uncertainty and confusion are, in fact, built into the system."[18] Reider notes that the ecologists and engineers "seemed to be grappling with unsettled questions inherent in the contemporary concept of nature itself."[19] Thus, one model—a simpler, more mechanical one, perhaps amenable to a cabin ecology—confronts another, far less determinate model, destabilizing categorical assessments of Biosphere 2's environmental logic.

Biosphere 2 seems therefore to have produced two overlapping logics. At one level the creation of wild biomes established representations of commonly identifiable "earthlike" ecosystems, both imported and constructed. At a functional level, however, they produced an experimental performative and durational system, with no guarantee that a recognizable image of the Earth in miniature would prevail. Would it become "urban weed," as ecologist Lynn Margulis predicted—a kind of "synthetic ecosystem" mush?[20] Having overloaded the biomes with species of flora and fauna, and having imported intact sections of natural biota, the Biosphere 2 researchers and crew were entering into a closed-system scenario in which they could not predict every element of the contents in advance. Instead the project would be directed toward measuring and monitoring what emerged, both shaping the direction of adaptation and in turn being shaped by the results.

This experimental, durational quality of the Biosphere project finds perhaps a more productive resonance with the kinds of conceptual problems confronted in the field of cybernetics than in cabin ecology. In his history of early British cybernetics, Andrew Pickering highlights the particular "ontology of unknowability" that these experiments embraced.[21] In order to construct responsive and adaptive machines, cybernetics engaged what Pickering describes as *the entrainment of the agency of nature*, intact into a human project, without penetrating it by knowledge and reforming it."[22]

For cybernetics this meant that the machines and systems such as Grey Walter's "Robot Tortoise" or Ross Ashby's "Homeostat" responded to environmental stimuli through simple but adaptive circuitry, where the behavior depended entirely on the machines' performative interaction with the world, and not on any stored information. The environments explored by these machines remained simultaneously opaque and constitutive to the cybernetic system. As Pickering puts it, "their worlds remained unknowable Black Boxes to the machines."[23] In this sense, we might understand the imported chunks of coral reef or the microbial soups circulating through the mangrove swamps in Biosphere 2 in a similar way—as "black boxes"—not meticulously assembled simulacra but chaotic entities that nonetheless performed complex ecosystemic functions.

Laboratory Living

The wild biomes were not the only emergent, performative systems in Biosphere 2. The manned mission, as we have suggested, was a constitutive ingredient in the ecotechnics of the closed system, and one that occupied a particular niche within the structure. The Biosphere 2 enclosure was designed to accommodate an eight-person crew in what was called the "Human Habitat," or the "Micropolis"—25,000 square feet of living space, 240 feet wide and 87 feet high—consisting of studio apartments, laboratories, a machine shop, medical center, analytical lab, command room, kitchen, exercise facilities, library, observation deck, and lounge. The ground floor was occupied by domestic animals. Above them, the studio apartments measured approximately twelve by sixteen feet, with a mezzanine area bedroom overlooking a sitting room below. A veranda with a 160-foot-long row of vegetable planters (a "salad bar") ran the length of the apartments. The library and observation dome crowned the top of the Human Habitat. From here, the residents looked out over the agriculture biome below them, or onto the Arizona desert outside.

We should note that the Micropolis was imagined as a "city" biome, in which the eight human "biospherians" were regarded as technologically modern participants, rather than atavistic figures of a lost ecological balance. Furthermore, there was no "backstage" to the closed system: to be inside Biosphere 2 was to be a part of it completely. The city and agriculture biomes were fully within the enclosure and shared the same atmosphere as the rainforest, the desert, and the other "wild" biomes. Although the city and agriculture biomes represented the anthropogenic elements of the experiment and included many of the recognizable typologies of modern life, they also experimented with what an ecotechnic lifestyle might look like and consist of. Biosphere 2 problematized, rather than pastoralized, human-ecological relations within the experiment, insisting that they could consist not only of natural and biological but also technical, urban, and to a certain degree, synthetic elements.

One of the overlooked aspects of this arrangement is how the Human Habitat of Biosphere 2 (and perhaps the principles of ecotechnics in general) stepped outside the norms of the mainstream imagination of "futuristic" and technologically advanced dwelling. From space settlements to high-tech housing, Biosphere 2 could be seen as a showcase of potential applications for technologies, diets, routines, and forms of habitation. And yet there was no attempt within the Micropolis to preserve an image of the nuclear family in its vision of the ecotechnical future. Felicity Scott and others have pointed out that popular speculative proposals such as Gerard K. O'Neill's gravitational ring planet, immortalized in the drawings of Rick Guidice and Don Davis, depict

Figure 8.3
Human Habitat sectional diagram showing the location of the living quarters, kitchen, laboratory, recreation room, workshop, and library at the top. Drawing by the Institute of Ecotechnics, ca. 1991.

closed-system colonies of American libertarian suburbanism, nestled into a picturesque countryside, "liberated not only from gravity and friction but also from inhospitable climates, material scarcity, 'large scale governments,' and other Earthly threats."[24] Other canonical examples such as Peter and Alison Smithson's House of the Future for the Ideal Home Show in 1956 (or, for that matter, Monsanto's House of the Future at Disneyland in 1957), and even Street Farm's comparatively radical DIY (do it yourself) alternative, the EcoHouse of 1972, all take the single family home to be the basic unit of habitation. Biosphere 2 seems to ignore this protocol, instead suggesting a more ambiguous form of cohabitation (or co-evolution) typified by individual apartments, shared facilities, gender balance, and collective work and leisure routines. Not then a family, but a "crew"; not a house, but a habitat, with access to agriculture and "wilderness," absent the pretense of anthropological primitivism.

Biospheric Culture

The biospherian crew were engaged in all aspects of the daily running and maintenance of the enclosure. Although guided by the management of Space Biospheres Ventures from the offices of Mission Control, the crew made their own day-to-day schedule and established routines over the course of their two-year mission. According to biospherian Sally Silverstone, an average day in Biosphere 2 started with milking goats and collecting fruits and vegetables for the kitchen, followed by breakfast and a meeting with all members of the crew. After agreeing on daily goals and objectives, they would spend a number of hours tending to the 20,000-square-foot intensive agriculture biome and then split up to attend to various sections of the enclosure.[25] This could include anything from mechanical repairs to field studies, from weeding to tree surgery. Cooking was a rotational duty among the members, although Roy Walford, the nutritionist and medic in the group, closely monitored their diet and physiology.

Figure 8.4
Mission 1 crew, agricultural work: Taber in foreground, Jane near window, and Abigail and Mark van Thillo planting rice. Photograph by the Institute of Ecotechnics, 1991.

The intensity of a small group of individuals working together for an extended period of time played out in both psychological and cultural dimensions. As Mark Nelson recalls, the group engaged in a kind of self-analysis through a chapter-by-chapter reading of Tavistock psychologist W. R. Bion's *Working in Groups*.[26] At the beginning of the experiment they brought art supplies and musical instruments into the Human Habitat, which later combined with their communications equipment to form the basis of a music video.[27] The library was stocked with books on theater, ecology, ancient history, religion, science fiction, and practical DIY instruction manuals such as *How to Grow More Vegetables Than You Ever Thought Possible on Less Land Than You Can Imagine*."[28] The group used regular holidays and birthday celebrations to break up the monotony of their routine and they often ate meals in different parts of the biosphere. In many ways they tried to establish some kind of new normal that would be commensurate with the strangeness of their experience.

What is of particular interest here are the ways in which this human cultural and psychological experience of the closed system came to be entangled and conditioned by the workings of the various ecologies inside Biosphere 2. Aside from monitoring

Figure 8.5
Mission 1 crew parties in the Command Room. Photograph by Roy Walford, ca. 1991–1993.

their own basic physical health within the system, the biospherians also spent a good portion of their day maintaining the viability of the closed system itself. Knowing what to do often depended on knowing what to look for and how to measure the changes that were occurring in the various parts and processes of the systems. Biosphere 2 was outfitted with an array of over two thousand sensors, monitoring all the biomes in the enclosure, including atmospheric content, irrigation and moisture, light, temperature, and electronic and mechanical functions—a nerve system that conducted "a kind of Whole Earth CAT scan," as Curt Suplee joked.[29]

The biospherians were tasked with extensive field observations, tracking animal populations, and observing patterns in the growth and behavior of various plant species. These observations, along with the sensor data, which was collected and managed by a cluster of computers inside the "analytical lab" located within the Human Habitat, were analyzed and compared with data collected by Mission Control. The whole system created a detailed picture of the atmosphere, soil, and biological populations, so that the granular shifts and adaptations occurring within the enclosure could be studied, then action taken to push these factors in one direction or another. The construction of this nerve system was an explicitly cybernetic approach to the long-term human-ecological entanglement within Biosphere 2. The relationship of feedback and adaptation engaged both the biomes and the humans in a reciprocal process of communication. Referring to the apparatus of the nerve system, botanist Walter Adey, designer of the coral reef in Biosphere 2 said, "It's like building a computer model of the mind which then turns around and interrogates the mind that created it."[30]

The definition of what constituted sensors and data within Biosphere 2 was broad. The biospherians themselves were sensors and data, deployed throughout the enclosure to both measure and be measured. Likewise, John Allen noted that, "It took some time before the engineers realized that observations of indicator species—plant or animal—which were sensitive to changes in nutrients, acidity of soil, or water quality could be as accurate a sensor as an electronic device."[31] Through this simultaneous and overlapping application of natural and electronic sensors, Biosphere 2 engaged in a hybrid kind of environmental intelligence, something the cybernetician Stafford Beer might have called "biological computing." Perhaps most famous for his development of the Cybersyn project in Chile in the early 1970s, Beers had earlier put forward detailed proposals for various biological agents that might be used as a governing mechanism for a factory, including a pond ecosystem. As Pickering has pointed out, the crucial aspect of Beer's turn toward biological computing was that he had hit upon the fundamental intelligence of complex adaptive systems.[32]

Figure 8.6
Ecotechnics work: Taber McCollum collecting air samples. Photograph by the Institute of Ecotechnics, 1991–1993.

In Biosphere 2, this adaptive intelligence could be attributed to both the biospherians and their biomes in turn. As Kevin Kelly describes in *Out of Control*, "Within great latitudes, the artificial ecosystem of Bio2 [sic] ran its own course, but when it veered toward a runaway state, or stalled, the biospherians nudged it."[33] This was especially true when the environmental effects seemed to directly impact the biospherians' own niche in the ecosystem. A year into the experiment, oxygen and carbon dioxide levels were out of balance, and oxygen in particular was "disappearing" within the system. A 6 percent change in oxygen concentration, combined with excess carbon dioxide, created the atmospheric equivalent of a much higher altitude, giving the biospherians headaches, making them lethargic and short of breath. To act against these fluctuating gasses, they raised the heat in the savanna and desert, encouraging a spring bloom of leaf buds; they also employed a carbon scrubber device to sequester some of the excess carbon dioxide and injected pure oxygen into the system to address what later turned out to be a chemical reaction with exposed concrete, which was forming calcium carbonate as it dried, sequestering oxygen and carbon in the process.[34]

This palpable sense of a chaotic but essential integration of material and energetic cycles within the closed system created a powerful sense of connection for the biospherians. There could be no toxic chemical solutions, no waste that would not return in a different form, no substance that was not food or energy for another part of the system.[35] As philosopher Timothy Morton has reflected:

Two hundred years of idealism, two hundred years of seeing humans at the center of existence, and now the objects take revenge, terrifyingly huge, ancient, long-lived, threateningly minute, invading every cell in our body. When we flush the toilet, we imagine that the U-bend takes the waste away into some ontologically alien realm. Ecology is now beginning to tell us of something very different: a flattened world without ontological U-bends. A world in which there is no "away."[36]

The biospherians also spent increasing amounts of time responding to the signals and needs of the biomes, creating the necessary effects to sustain the ecosystems within their complex artificial terrain. They would "act out the larger forces of nature beyond the ecological community," as Adey put it.[37] By "larger forces of nature," Adey meant wind, waves in the ocean, sudden rainfall, and brush fires. All of these phenomena were replaced with technology in Biosphere 2: fans, a wave machine, sprinkler systems. But as Kelly suggests, even these were too predictable, and the biospherians had to be the agents of a generative turbulence, playing the role of stimulators, agitators, foresters, engineers, plumbers, and so on.[38] The intense, sometime unexpected and protracted physical labor of working on the biomes and in the farm also created the unintended consequence of needing more calories than were initially calculated.[39] In response, the biospherians tried to maximize the space allotted for food production, filling the stairwells of the Human Habitat with planters.[40] On a certain level, the humans came to realize that they worked *for* the Biosphere itself, sometimes struggling to keep up with the changes that were taking place. As Kelly puts it, "They shared control with the emergent system itself. They were copilots."[41] Similarly, in describing the orientation of biological computing, Pickering writes that it "entailed a much more *symmetric* relation between the human and the nonhuman–a 'conversation,' as [Gordon] Pask put it, a 'compromise,' in which human performances and goals, the specifics of management, were themselves liable to open-ended transformation—mangling—in negotiation with ponds or electrolytic cells, performative black boxes."[42]

The assemblage of biomes, the technosphere of monitoring equipment, fans, filters, and pumps, the Human Habitat—Biosphere 2 was less a model of the Earth and more a feral monstrosity, the ecological equivalent of Frankenstein's monster, which if anything suggested an irretrievable entanglement of humans, technology, and natural systems. Critics saw in Biosphere 2 a functionalist utopia/dystopia—a self-regulating machine that might justify deregulation experiments elsewhere in the culture.[43] However, the reality was far more chaotic and complex. Biosphere 2 transgressed the norms of both economic viability and institutional science. It required a patient and durational performance—one riddled with complex problems and improvised solutions, one that demanded the full attention of the human participants and an ever-increasing investment of money and resources.

Scientific Theater

The insistence on a durational closed system that included humans made it difficult for Biosphere 2 and its management to conform to the norms expected by both the media and the scientific establishment. Indeed, the ambitions of Biosphere 2 attracted critical attention before the structure of the building was even complete. In a moment of clear foresight, journalist Curt Suplee warned, "The whole venture is so confoundingly fantastic that it is certain to be enthusiastically misrepresented."[44] The relative anonymity surrounding Space Biospheres Ventures—the mysterious organization with deep roots in 1970s counterculture that conceived of the experiment[45]—and the elusiveness of its financier, Ed Bass, haunted the mass-media perception of the project.[46]

Initial points of contention for journalists covering the project were the credentials and qualifications of its founding members and management. "One of the things most obvious about the principles of the Biosphere project is a shortage of scientific degrees," wrote one reviewer of John Allen's book about Biosphere 2.[47] Only two out of eight members of the management team had graduate degrees in science. The press ignored or discounted the decades of self-directed learning, conferences and field research, as well as the consultation and endorsement of an international community of scientists that preceded Space Biospheres Ventures' production of Biosphere 2.[48] Instead, they focused on the group's background in experimental theater practice, often in a derogatory fashion. "In short, the Biospherians may be talking science, but what they are doing is more akin to well-financed science fiction."[49]

Ultimately, Ed Bass, who backed the project financially and had been involved with the Institute of Ecotechnics from the very beginning, began to doubt the profitability of an open-ended cybernetic exploration of human-environmental relations in a three-acre closed system over the course of 100 years. In order to ensure public confidence in the research and to assuage the press, a scientific advisory committee was formed. Members included high-profile consultants to Space Biospheres Ventures such as Tom Lovejoy from the Smithsonian, Gerald Soffen of NASA, Ghillean Prance from Kew Gardens, and Eugene Odum. As Reider observed, "The committee's aims were scientific but also theatrical: to improve the Biosphere's science program and to rebuild its reputation."[50] This committee was in many ways the beginning of the conceptual dismantling of Biosphere 2 (the social and physical dismantling would come later). The committee, although supportive in its initial resolutions, principally saw in the Biosphere 2 project an experimental university laboratory that needed only to be optimized for better amenability to research grants. They produced a report in 1992 that in many ways echoed the subtext of the conflict inherent in the previous battles over the site. "The committee is in agreement that the conception and construction of Biosphere 2 were acts of vision and courage," the report read, going on to suggest steps for scientific conventions and rigor by which the project was not abiding. The report sets "research priorities for ongoing and future projects" and calls for "detailed project budgets, states long- and short-term goals, and outlines methods, anticipated results, and potential significance."[51]

In the middle of Mission 1 the Biosphere 2 project was clearly split between two lines of thinking: those who believed in the durational performance of humans and nonhumans in a closed system, and those who thought it interfered with the work of a real laboratory.[52] Most significantly, the committee itself was split. "The mission of this venture is not generally understood by the scientific community," wrote Odum,[53] while Lovejoy complained "that the initial business about eight people being closed up for two years was so hyped that the science got lost."[54]

Questions like Traci Watson's—"Can Basic Research Ever Find a Good Home in Biosphere 2?"[55]—posed a conflict between different cultures of science. The controversy was, perhaps, less concerned with whether good or bad science was being done in the facility, and more with who was doing it. True to earlier countercultural endeavors in recreational drugs, architecture, education, farming, and business, Space Biospheres Ventures and the IoE pursued a sort of outlaw science. As one journalist put it: "They said they'd build this vast project, and they did it. It didn't take 20 years, as would a government-sponsored research project of this magnitude. They didn't have to spend half their time writing grants, they didn't have to proceed on a step-by-step, measured schedule of inquiry, they didn't have to have umpteen trial runs to make sure the basic scientific knowledge preceded the applied science—no, they just did it."[56]

Unable to agree or steer the project in a new direction, the scientific advisory committee dispersed. Hoping to end the conflict, Bass persuaded the U.S. District Court judge in his hometown of Fort Worth, Texas, that the management of Biosphere 2 was compromising the ecological integrity of the project. Armed U.S. marshals accompanied

Bass and his team when they assumed control of the site, while the other managers from Space Biospheres Ventures were campaigning in Japan and served with restraining orders. With its scientific merit in doubt, Space Biospheres Ventures was not able to sell its data or pursue grants available to the scientific community, though one possible spinoff technology materialized from the first two years of Mission 1—the Airtron, which took the form of an organic air purification system utilizing soil filtration for interior spaces. Space Biospheres Ventures hoped it would be the first of a series of new ecotechnologies. However, the majority of revenues came solely from tourists visiting Biosphere 2 and covered only a portion of the running costs.

After Life, Death

As the human experiment came to an abrupt end, Bass's newly appointed CEO, a former Goldman Sachs investment banker named Steve Bannon asked, "What was being gained by locking these people up for a year?"[57] Bannon's question—rehearsing a rhetorical skill he would use a quarter century later as the wildly controversial political strategist for Donald Trump—was echoed by the scientific community. Wallace S. Broecker, who was the head of a team from Columbia University's Lamont-Doherty Earth Observatory, referred to the project's initial goals of maintaining a closed-system experiment with a crew inside as "a stunt."[58] "It was interesting, but if we were to bind ourselves to those silly rules, the project would not be able to make the most of the structure's scientific potential," he said.[59] Under its new management by Columbia University from 2003 to 2007, and subsequently by Arizona University from 2011, Biosphere 2 moved away from modeling an adaptive closed-system world to being a straightforward vivarium laboratory. "It just needed to be cleaned out and recalibrated and then people could go in and find out what it could be used for," said Steve Bannon's brother Chris, who also became a long-time manager of the site for Columbia University and subsequently the University of Arizona.[60] In order to start anew, Columbia University went in and started cleaning house: "Giant fans borrowed from a local mining company," read one 1995 report, "are forcing warm Arizona desert air through the ersatz Eden, bringing carbon dioxide levels close to normal. Some 200,000 gallons of water have been replaced, as have 12 tons of soil."[61] Columbia "will also replace the 200,000 gallons of water that was recycled continuously during the project's first few years of use," reported another.[62] Other changes implemented were plastic sheet dividers that allowed the scientists to separate and close off the once interconnected biomes.[63] The agriculture biome was dug up and left barren. "We are, essentially, hitting the reset button," Chris Bannon announced.[64] Biosphere 2 would, in a sense, be open

Figure 8.7
View inside the Human Habitat packed up in boxes after an abrupt end to Mission 2. Photograph by the Center for Land Use Interpretation, ca. 1996.

for business. Literally opening the doors of the experiment completely transformed the nature of the experiment, and completely transformed the logic of the building and its intent. No longer utilized as a human habitat, the Micropolis was turned into a museum for visitors.

The view that these actions rescued Biosphere 2 proliferated in the press as a kind of blind trust that the University institutional system of science was going to produce the best possible results. With the humans gone, the system's biomes became mere experimental tools. The agriculture biome became the agro-forestry biome because it no longer needed to produce food. The ocean became known simply as a "tank." The desert biome, like the agriculture biome, did not prove to be a fruitful zone for funding, and was largely abandoned, furnished with a walkway for tourists. Like the culture of science in the world outside, the science inside Biosphere 2 became microdisciplinary, reductionist, and focused on economic sustainability. Experiments included one investigating the effects of global warming on piñon pine trees, and another ascertaining how much carbon dioxide could be absorbed by cottonwood—both experiments

Figure 8.8
Former intensive agriculture biome emptied under the management of Columbia University. Photograph by Ralo Mayer, 2009.

strongly appealing to the lumber and the fossil fuel industries. At one point, Columbia University displayed a Volvo automobile in the Human Habitat, creating a strangely accelerationist image of ecological catastrophe, ostensibly measuring the rate and the precise conditions by which carbon pollutants can deplete living systems. These studies implicitly isolated humans from being directly subject to the conditions being studied, making it impossible to ask how biospherics changed bodies and minds, and divorcing questions of cultural transformation from those of ecology.[65]

The radicality of the manned mission was its fundamental acknowledgment of the integral links between ecology and human experience. Biosphere 2's first crew had developed the concept of a "ninth biospherian," providing a vivid illustration of the ways in which the biospherians understood their own presence and ecological research inside the closed system.[66] Within the first year of the enclosure, the crew of Mission 1 lost weight very quickly—around ten to fifteen pounds per person—a cumulative weight of nearly another human body. As biospherian Mark Nelson mused:

We may have lost weight but that weight didn't disappear. All of those molecules: water, nutrients, oxygen, nitrogen, whatever, they were still in the biosphere with us. But instead of it being

in human form it was in the atmosphere, it was in the plant leaves, in mosquitos, and in our domestic animals, it was in the ocean water, it was still in there with us. It was like we had actually
organically contributed and we were floating around doing more than our normal biospherian
duties. We were maybe even absorbed into the concrete. We were part of the structure, the life,
and the texture of the entire Biosphere. It's now not a closed system … but during the period
when it was essentially a closed system which continued I guess for at least a year after that …
those molecules were still flying around in there. You know I still probably have a fair number of
molecules in Biosphere 2.[67]

Though Biosphere 2 can be related to many of the strands of mainstream twentieth-
century systems research, its eccentric, countercultural laboratory lifestyle superseded
the simple prometheanism that sees creativity and providence only in the human
actor. It can instead be compared to the "dark ecology" described by Timothy Morton—"an untotalizable non-whole, not-all set that defies holism and reductionism."[68]
The manned mission was an open-ended ontological performance of human and nonhuman worlds.

Notes

1. See Kevin Kelly, *Out of Control: The New Biology of Machines, Social Systems and the Economic
World* (Boston: Addison-Wesley), 1994; Rebecca Reider, *Dreaming the Biosphere: The Theater of All
Possibilities* (Albuquerque: University of New Mexico Press), 2009; Sabine Höhler, "The Environment as a Life Support System: The Case of Biosphere 2," *History and Technology* 26, no. 1 (March
2010): 39–58.

2. Recent artworks that engage Biosphere 2 include: Ben Rivers, *URTH* (2016), 19 min. Super
16mm film; Ralo Mayer, *frnknst9n Invocation (There's Something Lurking in the Space Between)*
(2013), 15 min. HD video; Alice Wang and Ben Tong, *Oracle* (2016), 9 min. 33 sec. HD video. See
also Fred Turner, *From Counterculture to Cyberculture: Stewart Brand, the Whole Earth Network, and
the Rise of Digital Utopianism* (Chicago: University of Chicago Press, 2006).

3. See Laurence R. Veysey, "New Mexico, 1971: Inside a 'New Age' Social Order," in *The Communal Experience: Anarchist and Mystical Communities in Twentieth Century America* (Chicago: University of Chicago Press, 1978), 279–406; Jean Baudrillard, "Maleficent Ecology," in *The Illusion of the
End* (Stanford, CA: Stanford University Press, 1992), 78–88; Adam Curtis, "How the 'Ecosystem'
Myth Has Been Used for Sinister Means," *The Guardian*, May 28, 2011, https://www.theguardian
.com/environment/2011/may/29/adam-curtis-ecosystems-tansley-smuts (accessed December 8,
2017). Curtis's account of Biosphere 2 is also cited as the source for the account given in Benjamin H. Bratton, *The Stack: On Software and Sovereignty* (Cambridge, MA: MIT Press, 2015), 315 and
352.

4. Timothy W. Luke, "Environmental Emulations: Terraforming Technologies and the Tourist
Trade at Biosphere 2," *Ecocritique: Contesting the Politics of Nature, Economy, and Culture* (Minneapolis: University of Minnesota Press, 1997), 95–114; Peder Anker, "The Ecological Colonization of

Space," *Environmental History* 10, no. 2 (2005): 239–268; Peder Anker, "The Closed World of Ecological Architecture," *Environmental History* 10, no. 5 (April 2005): 527–552.

5. See Marc Cooper, "Take This Terrarium and Shove It," *Village Voice*, April 2, 1991, 24; Joel Achenbach, "Biosphere 2: Bogus New World?," *Washington Post*, January 8, 1992, https://www .washingtonpost.com/archive/lifestyle/1992/01/08/biosphere-2-bogus-new-world/f2de366a -ed63-42a0-ae39-0393db18ea46/?utm_term=.d57653385355 (accessed December 20, 2017); Michael Zimmerman, "Review: Biosphere 2: Long on Hype, Short on Science," *Ecology* 73, no. 2 (April 1992): 713–714; Traci Watson, "Can Basic Research Ever Find a Good Home in Biosphere 2?," *Science* 259, no. 5102, March 19, 1993, http://science.sciencemag.org/content/259/5102/1688 (accessed December 9, 2017).

6. Cooper, "Take This Terrarium and Shove It."

7. Achenbach, "Biosphere 2: Bogus New World?"; Cooper, "Take This Terrarium and Shove It," 24; Zimmerman, "Review," 713–714.

8. John Allen, Tango Parrish, and Mark Nelson, "The Institute of Ecotechnics: An Institute Devoted to Developing the Discipline of Relating Technosphere to Biosphere," *The Environmentalist* 4 (1984): 206. "Ecotechnics" is also partly derived from Lewis Mumford's notion of "biotechnics," which is in turn an extension of the ideas of botanist and planner Patrick Geddes. Geddes develops the ideas of "paleotechnics" and "neotechnics" in his book *Cities in Evolution* (1915). See Patrick Geddes, *Cities in Evolution: An Introduction to the Town Planning Movement and to the Study of Civics* (London: Williams & Norgate, 1915), and Lewis Mumford, *Technics and Civilization* (Chicago: University of Chicago Press, 1934).

9. The Odums' work on closed systems for space travel is described in Anker, "Ecological Colonization of Space."

10. William Dempster and Mark Nelson, "Living in Space: Results from Biosphere 2's Initial Closure, an Early Testbed for Closed Ecological Systems of Mars," in *Strategies for Mars: A Guide to Human Exploration*, ed. Carol R. Stoker and Carter Emmart (San Diego, CA: American Astronautical Society, Univelt, 1996), 373–390.

11. Simon Sadler, "An Architecture of the Whole," *Journal of Architectural Education* 61, no. 4 (May 2008): 109; see too Simon Sadler, "The Bateson Building, Sacramento, California, 1977–81, and the Design of a New Age State," *Journal of the Society of Architectural Historians* 75, no. 4 (December 2016): 469–489.

12. Anker, "Ecological Colonization of Space," 259. See also Anker, "Closed World of Ecological Architecture."

13. Anker, "Ecological Colonization of Space," 239–240.

14. Ibid., 242, and note 13. This idea is also identical to the theories of Konstantin Tsiolkovsky (1857–1935), the cosmist philosopher whose work informed the early Soviet experiments in human physiology for space travel. See Konstantin Tsiolkovsky, *Beyond the Planet Earth* (Oxford, UK: Pergamon Press, 1960).

15. Anker, "Ecological Colonization of Space," 259.

16. Ibid., 243.

17. Reider, *Dreaming the Biosphere*, 78.

18. Curt Suplee, "Brave Small World," *Washington Post*, January 21, 1990, https://www.washing tonpost.com/archive/lifestyle/magazine/1990/01/21/brave-small-world/aadfcbf8-0b94-42c2 -81f4-5590e3236480/?utm_term=.8d0898db3e2b (accessed December 8, 2017).

19. Reider, *Dreaming the Biosphere*, 94.

20. Kelly, *Out of Control*, 137.

21. Andrew Pickering, *The Cybernetic Brain: Sketches of Another Future* (Chicago: University of Chicago Press, 2009), 23.

22. Andrew Pickering, "Beyond Design: Cybernetics, Biological Computers and Hylozoism," *Synthese* 168, no. 3 (June 2009): 472; emphasis in original.

23. Pickering, *Cybernetic Brain*, 27.

24. Felicity D. Scott, "Securing Adjustable Climate," in *Climates: Architecture and the Planetary Imaginary*, ed. James Graham, Caitlin Blanchfield, Alissa Anderson, Jordan H. Carver, and Jacob Moore (Zurich: Lars Müller, 2016), 90.

25. "People in Glass Houses Don't Discuss Sex: They Had Parties, TV, and Toilets, but No Way Out," *Independent*, November 1, 1993, www.independent.co.uk/life-style/people-in-glass-houses-dont-discuss-sex-they-had-parties-tv-and-toilets-but-no-way-out-sally-1501443.html   (accessed December 20, 2017).

26. Mark Nelson, *Pushing Our Limits* (prepress manuscript PDF, Tucson: University of Arizona Press, 2018), 195.

27. Ibid., 35. See also Dr. Roy Walford, *Ecological Thing* (1993), 5 min. 7 sec. music video, https://youtu.be/t4wRGyS2DG0 (accessed December 20, 2017).

28. Reider, *Dreaming the Biosphere*, 5.

29. Suplee, "Brave Small World"; John Allen, *Biosphere 2: The Human Experiment* (New York: Penguin Books, 1991), 120.

30. Quoted in Suplee, "Brave Small World."

31. Allen, *Human Experiment*, 124.

32. Pickering, *Beyond Design*, 478–479.

33. Kelly, *Out of Control*, 134.

34. J. P. Severinghaus, W. Broecker, W. Dempster, T. MacCallum, and M. Wahlen, "Oxygen Loss in Biosphere 2," *Eos, Transactions, American Geophysical Union* 75, no. 3 (1994): 33 and 35–37.

35. When cockroaches began to attack the food stores in the Human Habitat, the biospherians took turns patrolling the kitchen at night and vacuuming them up. They then fed the cockroaches to excited and confused chickens, which in turn provided protein for eggs collected in the mornings; see Nelson, *Pushing Our Limits*, 84.

36. Timothy Morton, "Architecture without Nature," *Tarp Architecture Manual: Not Nature* (May 2012), 24.

37. Kelly, *Out of Control*, 135.

38. Ibid.

39. Christopher Turner, "Ingestion: Planet in a Bottle," *Cabinet*. 41, (2011), http://www.cabinetmagazine.org/issues/41/turner.php (accessed December 20, 2017).

40. Nelson, *Pushing Our Limits*, 87–88.

41. Kelly, *Out of Control*, 134.

42. Pickering, *Beyond Design*, 485.

43. Luke, "Environmental Emulations," 95–114.

44. Suplee, "Brave Small World."

45. Veysey, "New Mexico, 1971."

46. William J. Broad, "As Biosphere Is Sealed, Its Patron Reflects on Life," *New York Times*, September 24, 1991, http://www.nytimes.com/1991/09/24/science/as-biosphere-is-sealed-its-patron-reflects-on-life.html?pagewanted=all (accessed December 7, 2017).

47. Zimmerman, "Review."

48. Howard T. Odum, "Scales of Ecological Engineering," *Ecological Engineering* 13, nos. 1–4 (June 1999): 7–19.

49. Cooper, "Take This Terrarium and Shove It."

50. Reider, *Dreaming the Biosphere*, 200.

51. Biosphere 2 Scientific Advisory Committee, "Report to the Chairman, Space Biosphere Ventures," July 1992.

52. Erik Carver and Janette Kim, "Crisis in Crisis: Biosphere 2's Contested Ecologies," *Volume 20: Storytelling*, 2009, http://c-lab.columbia.edu/0167.html (accessed July 28, 2017).

53. Eugene P. Odum, "A New Kind of Science," *Science* 260, 5110 (May 14, 1993): 878–879.

54. "Report Urges Biosphere 2 to Spend More Time on Science to Succeed," *New York Times*, July 26, 1992, http://www.nytimes.com/1992/07/26/us/report-urges-biosphere-2-to-spend-more-time-on-science-to-succeed.html (accessed December 9, 2017).

55. Watson, "Can Basic Research Ever Find a Good Home in Biosphere 2?"

56. Achenbach, "Biosphere 2: Bogus New World?"

57. Quoted in Reider, *Dreaming the Biosphere*, 218.

58. Quoted in Roy L. Walford, "Biosphere 2 as Voyage of Discovery: The Serendipity from Inside," *BioScience* 52, no. 3 (March 2002): 259–263.

59. Dan Sorenson, "Biosphere 2 Changes Merely 'Hitting Reset,' New CEO Says," *Tucson Citizen*, March 16, 1995, http://tucsoncitizen.com/morgue2/1995/03/16/109370-biosphere-2 -chang (accessed May 16, 2017).

60. Quoted in Reider, *Dreaming the Biosphere*, 220–221.

61. Tim Beardsley, "Down to Earth: Biosphere 2 Tries to Get Real," *Scientific American*, (August 1995): 24–26.

62. Sorenson, "Biosphere 2 Changes."

63. William J. Broad, "Paradise Lost: Biosphere Retooled as Atmospheric Nightmare," *New York Times*, November 19, 1996, http://www.nytimes.com/1996/11/19/science/paradise-lost-biosphere -retooled-as-atmospheric-nightmare.html (accessed December 9, 2017).

64. Sorenson, "Biosphere 2 Changes."

65. Again, linking this to Pickering's account of cybernetics: "Second-order cybernetics, that is, seeks to recognize that the scientific observer is part of the system to be studied, and this in turn leads to a recognition that the observer is *situated* and sees the world from a certain perspective, rather than achieving a detached and omniscient 'view from nowhere.'" In Pickering, *Cybernetic Brain*, 25–26.

66. Ninth biospherian is first mentioned in Walford, "Biosphere 2 as Voyage of Discovery," 262; ninth biospherian phenomenon is described in Nelson, *Pushing Our Limits*, 5, 214.

67. Mark Nelson, interview by Ralo Mayer in *The Myth of The Ninth Biospherian/Biosphere 2 as Frankenstein's Monster* (2012), 7 min. 13 sec. HD video, transcript by author.

68. Timothy Morton, "This Is Not My Beautiful Biosphere," in *A Cultural History of Climate Change*, ed. Tom Bristow and Thomas H. Ford (London, UK: Routledge, 2016), 236.

9   Science Facts and Space Fictions: Making Room for the Frontiersman, Soldier, and Scientist in the Space Laboratory

Nicole Sully, William Taylor, and Sean O'Halloran

I have sought to offer humanists a detailed analysis of a technology sufficiently magnificent and spiritual to convince them that the machines by which they are surrounded are cultural artifacts worthy of their attention and respect.[1]

On November 4, 2011, a multinational crew comprising six "marsonauts" disembarked with considerable fanfare from a tin shed attached to the Russian Academy of Sciences' Institute of Biomedical Problems (IBMP). Like characters in the opening line of a comic routine, three Russians, a Frenchman, an Italian, and a Chinese volunteer had been confined in the structure for 520 days, the final period of a three-phase isolation experiment known as the Mars500 mission. This was a research project designed to anticipate the duration and experiences of a spaceflight to the Red Planet. It was a hypothetical journey that aspired to be explorative (to go "where no man" had gone before as *Star Trek* proposes), as well as diplomatic and scientific in conduct. The mission's expressed scientific purpose was to test the social, physiological, and psychological toll of space travel in three mocked-up environs replicating the confined spaces of an Earth-to-Mars spacecraft, a planetary ascent and descent vehicle, and a simulacrum of the Martian surface.

Like earlier settings forming the genealogy of environmental reasoning, the Mars500 space laboratory mock-ups were "analogues" of nature where human behavior could be observed in its spatial context in a heightened manner. However, unlike the analogues of history—the tropical island Edens of European colonialism, or the Victorian-era glasshouses and topographical fictions of the same period—the "nature" mimicked at the IBMP was extraterrestrial. The mix of fact and fantasy prevailing there (its "otherness" one could say) was further intensified.[2] For the duration of the trial, crewmembers were observed at extremely close quarters, in accommodation deprived of daylight and the additional stimuli that enrich the earthly environment. They were subjected in real time to the successive stages, necessary disciplines, and scientific protocols, imagined for an interplanetary journey. Supervening upon this scientific and wholly

Figure 9.1
Simulated spacewalk during Mars500 Mission, February 2011. *Source*: ESA/Mars500.

speculative enterprise, experiments presupposed a complex of interconnected subjectivities conditioned by a (very) long spaceflight. This was a journey in which each of
the five crewmembers were pressed into service as explorers, diplomats, scientists and,
simultaneously, research "guinea pigs."

Generally speaking, space laboratories, including historic ones such as NASA's Skylab (1973–1979), the International Space Station (from 1998 onwards) and the IBMP
facility, as well as fictional examples, such as the spacelab greenhouses from the 1972
sci-fi film *Silent Running*, demonstrate the Janus-faced character of the humanist subject
very clearly. In these cases, thinking about human life in space became the means for
understanding the requisites of humanity on Earth. The occupants of the extraterrestrial craft were observers of organic and inorganic processes contained by experiments,
packaged on board the spaceships, *and* living beings whose sociology, physiology, and
psychology were equally governed by the same experimental envelope. The vessels
demonstrate the bifurcated contours of what Bruno Latour studied as "laboratory life"
whereby scientists "make" scientific knowledge, but they are then, in turn, constructed
as members of a distinctive scientific culture, akin to members of an exotic tribe

whose domiciles, tools, and rituals become too useful, too necessary, and too costly to put aside.[3]

Vaguely mirroring and fusing these subject positions (the explorer, the diplomat, and the scientist) is another biographical figure in the character or "personae" of the marsonaut; kin to the astronaut and cosmonaut. Like a funhouse mirror that twists and distorts what it reflects, the life stories and lifestyles of space travelers, their machismo, bravery, and postorbital letdown, are the stuff of legend, as exemplified by Tom Wolfe's 1979 book on astronauts and test pilots, *The Right Stuff*, as well as occasional parody. Again, there are lessons from the history of environmental reasoning relevant to understanding the space traveler's lifestyle in this way. Human understanding of nature has obviously changed throughout the history of Western thought. However varied its forms and forces may have appeared in the past, there persists a long-standing view of the distinctiveness of our habitation of the earth. This view clearly prompts much environmental thinking as well as philosophical speculation, including Heidegger's existential musings in his 1951 essay "Building Dwelling Thinking" written just before the race to space took off.[4] The idea of habitable space and the figure of the *inhabitant* of both natural and built environments is an important, though easily overlooked, focus of theory and practice, history and present concerns. This figure entails a view of human beings as uniquely suited to their surroundings as members of other species are to theirs. So, what does it mean to inhabit space? How does one prepare for such a life, live through it, and then experience its legacy afterward? What are the varieties of space travelers? There is a rich trove of narratives, historical and fictional, to frame responses to these questions, although the answers are likely to be equivocal. Representations of space travel contribute to a long history of blending science fact and science fiction that has been a critical part of advancing knowledge of space— invoking a combination of sustained imagining, scientific experiments, and narrative accounts offered in literature and cinema, such as those that form the subject of this discussion.

This chapter considers the interplay between science and fiction in the history of space exploration, and in particular the journey to realizing the long-dreamed-of laboratory in space. It begins by examining early accounts that fuse science and fiction as part of the act of imagining travel to other worlds and the experience of inhabiting their terrains. It will then consider a number of cases in the history of real and imagined space travel that progress toward, and eventually realize, the "space laboratory." The chapter's conclusion returns to the Mars500 mission and the reception of its crewmembers once they left their cramped little shed by means of enlarging a historical and theoretical understanding of the lifestyle of the space traveler.

Early Science Facts, Early Space Fictions

The history of space fiction, like the history of space travel itself, has often been constructed around accounts of firsts, innovations, and milestones. This is a case not only of narrative technique, but also of the technological innovations imagined by the authors.[5] To begin with, many of the earliest narratives conformed to the characteristics established in early utopian fiction—with second-hand accounts of stumbling upon unfamiliar societies, which had developed as a microcosm, isolated from civilization. Second, they frequently resembled travelogues and travel literature, in particular the castaway genre, that grew in popularity during the eighteenth and nineteenth centuries, as scientists and explorers were themselves taking to the seas to discover the new world. The castaway genre often frequently reflected as much on the author's own civilization as those that had been discovered, with travel invoked as metaphor for a journey to the self. In the twentieth century, these literary models would become the blueprint for the weekly television adventures of, among others, Captain Kirk and the Starship Enterprise in *Star Trek* (1966–1969), and the Robinsons, marooned on Alpha Centauri, in *Lost in Space* (1966–1968), which both premiered at the height of the space race. These fictional journeys of space travel and space colonization have frequently been the grounds for testing and circulating ideas—both scientifically and philosophically—that have helped influence and inspire the scientists who would try to match these achievements in reality.

Like its fictional counterparts, accounts of the history of space travel have commonly entailed a mixture of historical fact, alongside rhetorical narratives. Further, the experiments and studies that made manned space flights possible—such as the astronaut training procedures described by Mary Roach in *Packing for Mars*, or the Mars500 experiments described earlier—involve simulations and role-playing that in themselves entail acts of imagined travel and vicarious habitation, albeit under controlled conditions.[6] The history of space exploration, particularly in its early years, was defined by national rivalries, milestones met and lost, and the battle to save "free minds," in a structure seemingly borrowed from narrative fiction and cinema. Scientists and politicians yearned to be the first to match the milestones already achieved in science fiction—orbiting in space, a lunar landing, manned planetary flights, and establishing a space station. The latter would be part laboratory, part residence, where the scientists, like the travelers in space fiction, would become both the scientific facilitator and subject, within a closed ecological system. The primary motivation for such environments as commonly stated in the mission statements of space agencies is less about the discovery of new and unfamiliar environments and more about being better able to understand humans and Earth.

Among the earliest examples of literary imaginings of travel to "other worlds" were satirical, philosophical, and later, scientific accounts. Lucian of Samosata's satirical *True History* and Plutarch's "Of The Face Appearing within the Orb of the Moon," both composed in the first two centuries AD, are believed to contain the earliest accounts of travel to other worlds, as well as descriptions of what might be encountered. Both works were known and read during the Renaissance, when imagined accounts of space travel first began to combine narrative fiction with scientific speculation. Instrumental in this early fusion of science and fiction was the famed German astrologer and astronomer Johannes Kepler, widely regarded as one of the most important figures in the scientific revolution.

In 1608, one year before formulating his laws of planetary motion that would influence understandings of astronomy and scientific thinking for centuries to come, Kepler authored one of the earliest works of science fiction, a novella titled *Somnium* that described an imagined journey to the moon.[7] *Somnium* was a curious mix of scientific theories and narrative fiction that anticipated many meditations on the theme in the centuries that followed. As Roger Launius observed in his essay "Prelude to the Space Age," Kepler's account includes speculation "on the difficulties of overcoming the Earth's gravitational field, the nature of the elliptical paths of planets, the problems of maintaining life in the vacuum of space, and the geographical features of the Moon."[8] In 1980, Carl Sagan suggested that Kepler deliberately engaged with fiction as a means to "explain and popularize science."[9] For Kepler, embedding scientific theory within a fictional premise likely provided a safeguard in politically volatile times—to distance the author from some of the more controversial aspects of his scientific theories—as well as providing an opportunity to indulge in speculation about the nature of the lunar landscape that would, perhaps, seem out of place in a scientific treatise.[10] Like many of the early works of space fiction, Kepler pondered possible ways to undertake such a journey. In *Somnium*, the journey would be facilitated by a combination of opiates and witchcraft. Although taking only four hours in duration, lunar travel, as it was described, was only possible during a solar eclipse, incredibly dangerous, and unsuited for the many. He wrote:

We do not admit desk-bound humans into these ranks, nor the fat, nor the foppish. But we choose those who regularly spend their time hunting with swift horses, or those who voyage in ships to the Indies, and are accustomed to living on hard bread, garlic, dried fish and other abhorrent foods.

The best adapted for the journey are dried-out old women, since from youth they are accustomed to riding goats at night, or pitchforks, or travelling the wide expanses of the earth in
worn-out clothes. There are none in Germany who are suitable, but the dry bodies of Spaniards
are not rejected.[11]

Despite the supernaturally assisted and narcotics enhanced travel, Kepler's scientific
inclinations would again prevail, when two years later in a letter to Galileo, he wrote:

It is not improbable … that there are in-habitants not only on the moon but on Jupiter too. …
But as soon as somebody demonstrates the art of flying, settlers from our species of man will not
be lacking. … Given ships or sails adapted to the breezes of heaven, there will be those who will
not shrink from even that vast expanse. Therefore, for the sake of those who, as it were, will pres-
ently be on hand to attempt this voyage, let us establish the astronomy, Galileo, you of Jupiter,
and me of the moon.[12]

While *Somnium* focused on travel toward, and a brief visit to, the moon, Kepler's
thoughts would eventually turn to the possibilities of inhabiting the lunar landscape.
In 1623, while revising the original manuscript for *Somnium,* he mused on the possibility of developing a *City of the Moon* work as a counterpoint to Campanella's utopian
City of the Sun (1602).[13]

As fantastical as Kepler's schemes were, such accounts were frequently used to lay
out scientific theories. Early narratives in the genre of space fiction consistently imagined scientific firsts centuries before they would become realities. For example, as Launius noted, the first fictional traveler to journey to the moon using a rocket was in
1649, in Cyrano de Bergerac's *Voyage to the Moon*, utilizing a method that anticipated
Newton's third law of gravity.[14] By the eighteenth and nineteenth centuries imagined
anecdotes of space travel were to become relatively commonplace, in particular, imagined journeys to the moon and Mars. Such work capitalized on the increasing popularity of travelogues, travel literature, and the "castaway" narrative, describing fantastical
visions of accidentally discovered worlds, frequently with scientific or philosophical
subtexts. Beyond simply traveling to the moon, writers began to imagine the possibilities of encountering built works, alien civilizations, and colonizing these landscapes
themselves.

In 1827, George Tucker's satirical *Voyage to the Moon*, saw space travelers utilize an
airtight, copper vessel with sliding doors and a primitive form of mechanical ventilation. Whereas Keppler's supernatural journey could be completed in four hours, Tucker's took three days. Upon arrival Tucker describes the territory of Morosofia and its
principal city, Alamatua which, like the cities known on Earth, included public buildings, shops, and streets. Alamatua contained approximately two thousand houses that
were constructed from "soft shining stone" with "porticoes, piazzas, and verandas,

suited to the tropical climate of Morosofia." The native inhabitants had also built "hot-houses," which he suggested commonly featured mirrored ceilings that "at once reflected the street passengers to those who were on the floor, and enabled the ostentatious to display to the public eye the decorations of their tables, whenever they gave a sumptuous feast."[15] Alamatua's hothouses were, likely, the first literary imagining of the possibilities of an artificial environment for the extraterrestrial cultivation of food, one that would later feature in cinema, such as the greenhouses that served as arks in *Silent Running,* and prefigure the visions of architects and societies to come, particularly those that imagine the space laboratory.

While Tucker's account described a civilization built by the moon's native inhabitants, other writers began to imagine the possibilities of building on, cultivating, and colonizing the moon. Among the earliest of these was Edward Everett Hale's *The Brick Moon* (1869–1870)—a serialized story that involved both space travel and lunar colonization. The work is notable, as Launius identifies, for entailing "the first known proposal for an orbital satellite around the Earth."[16] Like Kepler's fusing of science and fiction, Hale's narrative is interspersed with laborious scientific calculations and reasoning, and tediously detailed accounts of the vessel's construction. The final form, a brick moon, in its appearance was like that of "conglobated bubbles undissolved."[17] Numerous subsequent illustrations of this vessel consistently show a structure that closely resembles the spherical domes of eighteenth-century architects Claude-Nicolas Ledoux and Étienne-Louis Boullée.

As Nancy James notes, "fiction's first artificial satellite becomes the first manned satellite" when the structure is inadvertently launched into the sky with thirty-seven temporary residents (who had been seeking refuge from the winter within the warm spheres) still inside—to be lost for a year.[18] In Hale's story, the first structure on the moon is the vessel that had transported the characters there. As was the convention of the castaway genre, a series of convenient coincidences and accommodating conditions allow the inhabitants to not only survive, but to also cultivate their own sources of food and water. As with *Somnium,* the moon's climate is slightly tropical, yielding improved crop fertility, and prompting the basic building blocks the characters have at hand, such as lichens and chickens, to evolve into more productive and exotic species such as palms and hemlock, dodos and ostriches. As James writes, "The entire surface of the Moon is available for such activities as farming and outdoor recreation since the inhabitants still live inside the spheres."[19] Marooned on the moon, in the brick moon, a hermetic microcosm of society emerges that fuses the traditions of both castaway narratives and utopian fiction.

In *The Brick Moon*, the lunar surface is (as we now know) improbably farmed, its tropical and fertile conditions anticipate future concepts for the cultivation of space environments utilizing greenhouses. At this time, schemes for imagined space travel and the colonization of the moon began to move from literary exposition to scientific discussions. Thirty-five years after Hale's serialized novella, the idea of an orbital spaceship was taken up by a Russian school teacher named Konstantin Tsiolkonvsky who, like many of his predecessors, authored both science fiction and speculative science. His proposal for an orbital spaceship was published in a Russian science journal. One of the unique features of Tsiolkonvsky's scheme is the inclusion of a "space greenhouse" that, as Donna Goodman notes, "formed part of an enclosed ecological system."[20] The enclosed ecological system was to become the hallmark of twentieth-century schemes for space colonization. The hothouses of *The Brick Moon* and Tsiolkonvsky, while principally included in their narratives as a means to cultivate food in the context of long-term space colonization, anticipate later scientific research laboratories.

At around this time, in the 1920s, the first serious proposals for space stations begin to emerge. Among them, Hermann Oberth proposed a scheme for communications, the scientific observation of weather, and a means of enabling the refueling of space vehicles.[21] In his 1920 book *The Problem of Space Flight*, Hermann Noordung suggested a doughnut-shaped structure for scientific observation of Earth that was to be of sufficient size and density as to generate an artificial gravity.[22] The proposal included living quarters and an observation deck, as well as means to generate its own power.[23]

In the 1940s, proposals continued to be made with increasing seriousness, particularly as the Douglas Aircraft company began undertaking studies toward developing space technologies.[24] During the 1950s, discussion of the potential to colonize space moved from scientific discourse and science fiction, into the mainstream media. Seemingly impatient with the slow progress of real space exploration in comparison with its fictional counterparts, in March 1952, *Collier's Weekly* published a space-themed issue titled "What are we waiting for?" Among the various ideas discussed in the issue was a future space station that would be a functioning residential laboratory. In an article titled "Crossing the last Frontier," Wernher von Braun wrote: "Scientists and engineers now know how to build a station in space that would circle the earth 1,075 miles up. The job would take 10 years, and cost twice as much as the atom bomb. If we do it, we cannot only preserve the peace but we can take a long step toward uniting mankind."[25]

The promise of a future space station was also elaborated in Willy Ley's article in the same issue, titled "A Station in Space," which promised a "self-contained community, this outpost in the sky will provide all of man's needs, from air conditioning to artificial gravity." Ley continued, "Life will be cramped and complicated for space dwellers; they

Figure 9.2
Hermann Noordung, Three-Unit Space Station Concept, 1929. *Source:* NASA.

will exist under conditions comparable to those on a modern submarine."[26] The essay proceeded to lay out the details of the purpose of the space station, which was primarily to be focused on scientific endeavor and mechanical systems, as well as other features that would support its the functionality—ranging from sleeping quarters through to observation and communication equipment. Ley envisioned an interconnected network "with rocket ships in space, and with the space taxis that carry men from rocket ship to space station."[27]

It was during this time that the space race officially began. Conversations shifted from scientific speculation and literary imaginings into the realm of the public and politics. Perhaps the greatest advancement toward realizing a laboratory in space came on April 20, 1961, eight days after the Russian cosmonaut Yuri Gagarin became the first human to travel into outer space. U.S. President John F. Kennedy sent a panicked memorandum to his vice president, Lyndon Johnson, requesting an urgent report on the state of the space program. In the note that proved to be the catalyst for the manned Apollo missions to the moon, the president inquired, "Do we have a chance of beating the Russians by putting a Laboratory in Space, or by a trip around the moon, or by a rocket to land on the moon and back with a man. Is there any other space program which promises dramatic results in which we could win?"[28]

The perceived urgency of an American response to the latest in a series of milestones achieved by the Soviets was perhaps best indicated by the third question on the memorandum. Here Kennedy inquired, "Are we working 24 hrs a day on existing programs? If not, why not?"[29] Fueled by Cold War anxieties, and beaten by the Soviets to the most significant milestones in the race to space, the president's plea was ambivalent as to whether the most pressing goal for his country was to be the extraterrestrial conquest or the promotion of U.S. science, though clearly the one was entailed in the other.

The following month, Kennedy addressed Congress in the hope of obtaining additional funding to further the American space efforts. This address, taking place a few days after Alan Shephard became the first American in space, had a more measured tone. Having so recently experienced the taste of triumph, the president hoped to capitalize on the event's energy and excitement in order to secure the necessary funding to meet America's space goals. The ultimate goal was "to win the battle of men's minds." Kennedy's polished rhetoric became a measured call for "the new frontier of human adventure," a "great new American enterprise," and a gesture of leadership.[30] Public discussion of the space race, at this time, was predominantly focused on the "Race with the Russians." These two documents highlight the early tensions between exploration and science, politics and propaganda that characterized the first fifty years of the space program. As such, the early forays into space were presented as acts of *frontiersmanship*, undertaken largely by military personnel, rather than exploration by scientists.[31]

While matching the Russian milestone was an important coup, it would be important that America reach the next milestone first. Robert C. Seamans Jr., then the associate administrator of NASA, later recounted a meeting that took place the day after the Shephard flight, whereby

we all agreed that the fourth major reason for space, namely national prestige or national security in a broad sense, had been overlooked, had been misunderstood, and that we had to do something that was significant and appeared to be significant on a world-wide basis. … I remember that the Secretary [of Defense, Robert McNamara] wondered if the lunar landing was a big enough jump, feeling that the Soviets were possibly far enough ahead that they would get there before we did, even if we made a very intensive effort. He wondered if we shouldn't be considering manned planetary flight.[32]

The president's April 20 memorandum was by no means the first reference to a laboratory in space; rather, historical documents show that this had been one of the less publicized long-term goals of the space program since the 1950s.[33] A 1962 document outlining the objectives of the space program, identified the three key goals:

1. To provide insurance for the United States that its science and technology will not become obsolete in an age of explosive advances in scientific knowledge, engineering techniques and other technological innovation
2. To provide insurance against the hazards of military surprise in space technology and the possible adverse psycho-political consequences
3. To lead the world in the space age in such a manner as to derive benefits for all mankind.[34]

Thus, even at this stage, science was clearly seen as a means to preserve America's position. Among its principal assumptions, the document states that a lunar base was to be established within five to seven years of achieving a lunar landing. The mission target for 1966 included a "Three-Man Orbiting Laboratory," with the aim of achieving capacity to accommodate twelve men between 1970 and 1975.[35] In April 1963, Marvin Miles of the *Los Angeles Times* reported that NASA was seeking expressions of interest for developing a space laboratory. The paper stressed that project approval had yet to be granted, qualifying that the call was a reflection of their "continuing interest in the project." Outlining the benefit, Marvin wrote, "Crew members themselves would be subjects of experimental studies on human effectiveness and abilities in space."[36]

Seizing on the commercial possibilities offered by the anticipated space laboratory, in 1967 Barron Hilton (then chairman of the Hilton Corporation) speculated on the possibilities of an "Orbiter Hilton." In a 1967 address made to the American Astronautical Society, he described two prospects for his company's possible foray into space hotels, the first being the Lunar Hilton—which would be buried underground—and the second being the Orbiter Hilton. The latter, he speculated, was the most feasible given conversations he had undertaken with his friend Don Douglas Jr. (of the Douglas Aircraft), whose company, Hilton noted, had been exploring the concept of a space laboratory. It was anticipated that Hilton would lease space for its hotel from this laboratory, thus providing a temporary residence for scientists and other visitors.[37]

The shift in motivation from exploration and frontiersmanship to science was underscored on the very day of the first moon landing. On July 21, 1969, the *New York Times* published a long essay written by George E. Mueller, a NASA official, the title of which proclaimed "In the Next Decade: A Lunar Base, Space Laboratories and a Shuttle Service."[38] The article lay out the technologies developed for the space program that had already been adapted and implemented for life on Earth. Mueller declared that a "large flexible space station" comprising modular components was a logical next step in order to exploit the possibilities of weightless environments for generating new knowledge for Earth.

In the following decade, real progress was made toward realizing a laboratory in space. Once again, the American efforts were beaten to key milestones by the Russians, who launched Saylut in 1971. The first American space station, Skylab Orbital Workshop, was not launched until 1973, and crashed to earth, landing near the remote town of Esperance, in Western Australia in 1979.[39] During its lifetime, the latter hosted three manned missions. In addition to enabling scientific experiments, Skylab offered the opportunity to explore ideas of comfort and habitability in space. Between 1967 and 1973, NASA engaged industrial designers Raymond Loewy and William Snaith as "habitability consultants" to undertake a series of studies for the design of space environments, to be implemented on Skylab. Their sketches, and subsequent full-scale mock-ups, show an organized imagining of the realities of space life, and the need to design new equipment and procedures to undertake the most basic functions in this new environment. The grueling nature of space habitation was to be tested to the limits on Skylab's final mission—when, after eighty-four days in space, the astronauts famously went on strike.[40]

In parallel, real progress in space exploration furnished a new range of speculative proposals for space stations, such as the space colonies outlined by Gerald O'Neill in *The High Frontier: Human Colonies in Space* of 1976, or the Spider Space Station concept of 1977, with its ring-form resembling Noordung's 1939 concept. The reality of the space laboratory was to prove vastly different from that established in the cinematic representations, which in many ways paved the way for their existence. The spacious, minimalist, modernist temples to science and human achievement represented in films

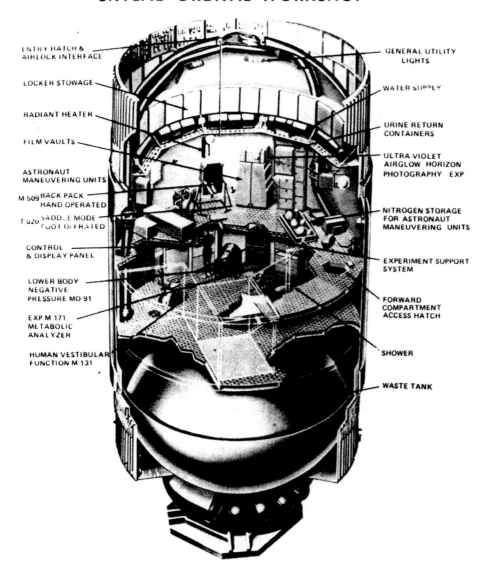

Figure 9.3a
(a) Cutaway view of Skylab Workshop, May 1973. *Source:* NASA.

Figure 9.3b
(b) Raymond Loewy and William Snaith, full-scale mock-up for a proposed artificial-G, shuttle-compatible space station. *Source:* NASA/MSFC.

such as Stanley Kubrick's *2001: A Space Odyssey* (1969) or, its ruined form, in Andrei Tarkovsky's *Solaris* (1972) were contemporaneous with the earliest venture of international space programs. Such depictions, through to, more recently, Ridley Scott's *The Martian* (2015), stand in contrast to the cluttered and cramped submarine-like existence of real space travel. In the "race to space" both the Americans and the Soviets were repeatedly outdone by the imagination and experiments of earthbound visionaries in the 1960s and 1970s, who pushed the limits of human habitat and science with experimental architecture and science fiction.

The focus in recent years has been on the building of the International Space Station, which commenced its piecemeal construction in 1998. During its construction, the astronauts were to add a new role to their existing ones of explorers, diplomats, and scientists—that of construction laborer. Permanently manned by an international

Figure 9.4
"Spider" space station concept, 1977. *Source:* NASA.

crew, ferried to and from by the Russian Soyuz spacecraft, the International Space Station, in uniting the key participants in the space race, was a symbolic end to the very international rivalries that made it possible.

Conclusion

Despite the seriousness of the program's aims and the personal sacrifice of its sequestered crewmembers, the Mars500 mission was widely mocked in the media. Before the marsonauts had even begun their extended stay in the IBMP shed, a headline in *The Guardian* newspaper on June 4, 2010, conveyed a sense of their defensiveness: "Mars mission in a Moscow hangar is no joke, say astronauts."[41] Another headline in the same newspaper in a January 2013 issue (well after their stay finished and the crew emerged exhausted and demoralized) read, "Fake mission to Mars leaves astronauts spaced out," continuing with the subheading, "Trip to Mars in pretend spaceship on Moscow

industrial estate affects sleep, activity levels and motivation of six-man crew."[42] Following decades of conspiracy theories claiming the U.S. Apollo 11 moon landing on July
20, 1969 had been faked, a simulated Mars mission in a Russian shed was an irresistible
target, particularly for the Western media about to enter what has since come to be
known as a "post-truth" age.

   Then again, the parodied treatment of the Mars500 mission was perhaps due in part
to the unconvincing aesthetics of its cramped laboratory, with its laminated wooden
flooring, faux-timber wall paneling, shelves of knick-knacks, and store-purchased furniture, including a Persian-style rug and floral cushions, particularly in comparison with
Loewy and Snaith's carefully curated Skylab mock-ups that included Saarinen's Tulip
Chairs, and a "model" styled with futuristic fashion, or the more sophisticated designs of
Biosphere 2 or the International Space Station (to say nothing of the unworldly Edenic
glasshouses in *Silent Running* [1972] and *Elysium* [2013]). Paradoxically, compared to
these monumental and aesthetically futuristic contraptions, the "shed" simply wasn't
believable. With its all-male crew, shabby and mismatching décor, and a steady stream
of photographs of breakfast-table high jinx in t-shirts and shorts, the crew too closely
resembled a household of students abroad on the Erasmus program. Such impressions
were not helped by an oft-reproduced image of a weightlessness simulation, which an
obscure caption on the ESA/Mars500 image database notes, was achieved as a result of
digital image manipulation, rather than scientific experimentation.[43]

   There may be additional reasons why such experiments in space travel and living are
suspect or incredible (for being unbelievable). As analogues—not so much of "nature"
as we know it on Earth, but rather of the terrestrial environment's boundless, vacuous,
and irradiated "other" of interplanetary space—the Mars500 spacelab and additional
examples of "true" space laboratories launched into the void or imagined floating
there in books and films are philosophically baffling as habitats. For one thing, the
highly artificial character of the spacelab environment challenges expectations and
circumstances required for "normal" science to proceed. Life on board carries heightened and potentially catastrophic risks arising from inadequate training, insufficient or
faulty technology, and a range of other factors that threaten the conduct of scientific
methods. Replacing the astronauts (of whatever ilk) with dogs, chimps, or robots may
seem to provide a solution, rendering the scientific enterprise on board more fully
closed and reliable, though the substitution merely removes the sociology one small
step (or giant leap, to quote Neil Armstrong) further, beyond the experimental envelope of the vessel to mission control, to Earth. Tellingly, no space film has robots that
wholly replace its human cast (perhaps only one or two has a robot as its hero and
then, accidentally), while characters like the psychopathic computer "Hal" in *2001: A*

Figure 9.5a, b
(a) Mars500 crew eating breakfast, June 2010. *Source:* ESA/Mars500 crew. (b) Mars500 crewmembers Romain, Yue, Diego, and Aleksandr preparing an ECG recording. *Source:* ESA/Mars500.

Figure 9.6
Mars500 crewmembers experience simulated weightlessness courtesy of digital manipulation, April 1, 2011. *Source:* ESA/Mars500.

Space Odyssey and corporate android operative "Ash" in *Alien* (1979) demonstrate how technological systems can be as corrupt, dysfunctional, or deceptive as their human makers. As Latour reminded us nearly forty years ago, his observation still valid today, science has established its unique cultural authority partly by excluding the sociology of laboratory life.[44] Thus, among the "experts" included among the crews of explorers, diplomats, and scientists in recent collaborative missions on board Mars500, Skylab, or the International Space Station (or in films like *The Martian* or *Life* [2017]), the roles of space sociologists and anthropologists have been left unfilled. In space films, while the tribal behavior of alien species may fascinate some crewmembers, no one seems to take much of a professional interest in the interpersonal dynamics governing their fellows. Such disinterest exacerbates the disorienting effect of mixed demands upon space travelers that they be adventurous, tactful, and reasonably objective all at the same time.

Just as the preceding observations reinforce the prevailing idea among philosophers of science that scientific enterprise does not occur in a (social) vacuum, the conduct of space travel is acutely vulnerable to factors "outside" the experimental envelope, including cultural and political pressures. While much valuable space research is now routinely conducted using remotely controlled interplanetary or interstellar probes, these fail to fully capture the public imagination or attract political patronage in the same way that the prospect of a manned mission to Mars or beyond does. As Kennedy noted in his famous "moon speech," delivered at Rice University, which supercharged the U.S.-Soviet space race, the allure of space exploration is in the challenges it poses.[45]

Having taking the American public on more than one far-flung ride, Kennedy's contemporary successor, President Donald Trump, has stated his goal to land Americans on Mars by the end of his second term, a clear rebuff to NASA's more modest plan to see humans on the Red Planet sometime after 2030.[46] Trump's chutzpah plays into the national mythology of the adventurous disposition and scientific superiority of the United States—at a time when these qualities are under question and scientific research budgets have diminished. Trump's mission relies on faith in the innateness of American leadership propagated in Hollywood disaster films from the 1970s—such as *The Poseidon Adventure* (1972) and *The Towering Inferno* (1974)—where salvation is delivered just in time by the few people standing tall above the crowd of willing followers whose authority has been sharpened by exceptionally audacious lifestyles and challenging life experiences. This is an ideological construct that has helped more than one U.S. astronaut find success on the American political scene following their retirement from space. John Glenn Jr., the first American to orbit the earth (February 1962), is the most well-known figure, but the list also includes Edwin Garn who flew on the space shuttle *Discovery* (April 1985) and Clarence Nelson who was an astronaut on the shuttle Columbia (January 1986). All three became U.S. senators. Conversely, faith in American leadership adds further poignancy to the biographies of former space travelers whose post-flight lives have been marred by depression and alcoholism, challenges documented, for instance, in the autobiography of former Apollo astronaut Edwin (Buzz) Aldrin Jr.[47]

Finally, the prospect of *human* life being contained and conveyed within the space laboratory carries a provocative charge for another reason. The prospect seems to heighten long-standing tensions that come from knowledge of our species' evolution and habitation of Earth. One of these appears in a narrative conflict commonly found in literature, particularly in fiction with an environmental or pseudo-ecological cast such as Victorian-era topographical literature. These are the stories about nineteenth-century Europeans who find themselves conveyed to remote and geo-graphically distinctive landscapes, such as the Scottish Highlands or the Swiss Alps, where they find themselves overwhelmed by the sublime. Or, it may be the exotic ter-rain of remote tropical islands where Europeans are shipwrecked and obliged to apply their characteristic skills and learning to the taming of "wild" nature. The conflict derives from two opposing perspectives on what makes humans distinct from nonhu-man animals. One sees humans as wholly determined by nature, incapable of escaping their genetic inheritance. The other sees humans as blank canvases shaped by cul-ture. Victorian writers, who liked to situate their characters' romantic distractedness in the tropical glasshouses of public gardens or big houses, invoked a dreamlike state to resolve the conflict. The physiological discomfort of the hot, humid environs of the glasshouse was momentarily reconciled with the protagonist's hot-blooded pas-sions by wistfully wondering "What if, if only Gerrard (or Hermione for that matter) were here?" In space movies, a comparable hiatus occurs when traumatized characters briefly reconnect via teleconference or voice message transmitted from loved ones on Earth: "What if, if only I were *not* here?"

Notes

1. Bruno Latour, *Aramis, or the Love of Technology*, trans. Catherine Porter (Cambridge, MA: Har-vard University Press, 1996), viii.

2. For discussion of the historic contexts for environmental reasoning see: Richard Grove, *Green Imperialism: Colonial Expansion, Tropical Island Edens, and the Origins of Environmentalism, 1600–1860* (Cambridge and Melbourne: Cambridge University Press, 1996); and William M. Taylor, *The Vital Landscape: Nature and the Built Environment in Nineteenth-Century Britain*, (Aldershot: Ashgate Publishing, 2004).

3. Bruno Latour and Steve Woolgar, *Laboratory Life: The Construction of Scientific Facts* (Princeton, NJ: Princeton University Press, 1986).

4. Martin Heidegger, "Building Dwelling Thinking" (1951), in *Poetry, Language, Thought*, trans. and introduction Albert Hofstadter (New York: Harper & Row, 1975), 145–161.

5. For an account of early works of space fiction see: Thomas D. Clareson, *Science Fiction Criticism: An Annotated Checklist* (Kent, OH: Kent State University Press, 1972); Arthur B. Evans, "The Ori-gins of Science Fiction Criticism: From Kepler to Wells," *Science Fiction Studies* 26, no. 2 (1999): 163–186; Roger D. Launius, "Prelude to the Space Age," in John M. Logsdon et al., *Exploring the Unknown: Selected Documents in the History of the U.S. Civilian Space Program*, vol. 1 (Washington, DC: National Aeronautics and Space Administration, NASA History Division, Office of Policy and Plans, 1998); David Cressy, "Early Modern Space Travel and the English Man in the Moon," *The American Historical Review* 111, no. 4 (October 2006): 961–982.

6. See Mary Roach, *Packing for Mars: The Curious Science of Life in the Void* (New York: W. W. Norton & Company, 2010).

7. The first two of Kepler's Laws of Planetary Motions were authored in 1609, the third in 1619. Carl Sagan, for example, in 1980 acknowledged Kepler's work as one of the first examples of science fiction writing in the television documentary *Cosmos: A Personal Journey*. Carl Sagan and Adrian Malone, *Cosmos*, collector's ed. (Studio City, CA: Cosmos Studios, 2000), episode 3.

8. Launius, "Prelude to the Space Age," 3.

9. Sagan and Malone, *Cosmos*, episode 3.

10. This was in fact a common literary device. Paul Turner, in his introduction to the 1965 Penguin edition of *Utopia,* for example, suggests a similar idea about the use of fictional premises for More's presentation of his ideal society. This idea is seemingly supported by Kepler's initial reluctance to publish the work until decades after it was written, at which stage the text was revised and expanded to include more scientific explanation. Ultimately, the work would not be published until after his death. That his scientific theories were contentious is supported by the fact that Kepler's mother was accused of witchcraft several years after the work was written (but before its publication).

11. Johannes Kepler, *Somnium (The Dream)*, 1608, *The Somnium Project*, trans. Tom Metcalf, 2011–2017, chapter 9, https://somniumproject.wordpress.com/somnium/ix/ (accessed July 28, 2017).

12. Johannes Kepler and Edward Rosen, *Kepler's Conversation with Galileo's Sidereal Messenger*, The Sources of Science, no. 5 (New York and London: Johnson Reprint Corp., 1965), 39, http://digitalcollections.library.cmu.edu/awweb/awarchive?type=file&item=393654 (accessed July 28, 2017).

13. On December 4, 1623, Kepler wrote in a letter to Matthias Bernegger, "Campanella wrote a *City of the Sun*. What about my writing a 'City of the Moon'? Would it not be excellent to describe the cyclopic mores of our time in vivid colors, but in doing so—to be on the safe side—to leave this earth and go to the moon." See Carola Baumgardt, *Johannes Kepler Life and Letters*, introduced by Albert Einstein (New York: Philosophical Library, 1951), 155–156.

14. Launius, "Prelude to the Space Age," 3.

15. George Tucker, *A Voyage to the Moon: With Some Account of the Manners and Customs, Science and Philosophy, of the People of Morosofia, and Other Lunarians* (New York: Elam Bliss, 1827), chapter VI, Project Gutenburg, http://www.gutenberg.org/cache/epub/10005/pg10005-images.html (accessed July 31, 2017).

16. Launius, "Prelude to the Space Age," 3.

17. Edward Everett Hale, "The Brick Moon, II," *The Atlantic Monthly* 24, no. 145 (November 1869): 603–611 and 607.

18. Nancy James, "Realism in Romance: A Critical Study of the Short Stories of Edward Everett Hale" (PhD diss., Pennsylvania State University, 1979), 132.

19. Ibid., 133.

20. Donna Goodman, *A History of the Future* (New York: Monacelli Press, 2008), 161.

21. Roland W. Newkirk, Ivan D. Ertel, and Courtney G. Brooks, *Skylab: A Chronology* (Washington, DC: Scientific and Technical Information Office, National Aeronautics and Space Administration, 1977), part I, https://history.nasa.gov/SP-4011/part1a.htm (accessed July 1, 2017).

22. Hermann Noordung (Hermann Potočnik), *The Problem of Space Travel: The Rocket Motor*, ed. Ernst Stuhlinger and J. D. Hunley with Jennifer Garland (Washington, DC: National Aeronautics and Space Administration, [1929] 1995), https://history.nasa.gov/SP-4026.pdf (accessed July 1, 2017).

23. Newkirk, Ertel, and Brooks, *Skylab*.

24. Ibid.

25. Werner von Braun, "Crossing the Last Frontier," *Colliers Weekly*, March 22, 1952, 24.

26. Willy Ley, "A Station in Space," *Colliers Weekly*, March 22, 1952, 30–31.

27. Ibid.

28. John F. Kennedy, "Memorandum for Vice President," April 20, 1961, Presidential Files, John F. Kennedy Presidential Library, Boston, https://history.nasa.gov/Apollomon/apollo1.pdf (accessed July 1, 2017).

29. Ibid.

30. John F. Kennedy, excerpts from "Urgent National Need," speech to a Joint Session of Congress, May 25, 1961, Presidential Files, John F. Kennedy Library, Boston, https://www.nasa.gov/pdf/59595main_jfk.speech.pdf (accessed July 1, 2017).

31. This language was remarkably similar to a 2004 Bush-era mission statement that later called for a renewed pioneering spirit.

32. Robert C. Seamans Jr. recorded interviews by Walter D. Sohier, Addison M. Rothrock, and Eugene M. Emme, on March 27, 1964, John F. Kennedy Library Oral History Program, 10, https://archive1.jfklibrary.org/JFKOH/Seamans,RobertC.,Jr/JFKOH-RCS-01/JFKOH-RCS-01-TR.pdf (accessed July 10, 2017). The conversation Seamans recounts took place on May 6.

33. For example, refer to Papers of John F. Kennedy, Presidential Papers, National Security Files, Subjects, Space activities: Long Range Plans of NASA (National Aeronautics and Space Administration), vol. I–IV, May 29, 1962, 10–11, https://www.jfklibrary.org/Asset-Viewer/Archives/JFKNSF-307-002.aspx. (accessed July 10, 2017).

34. Ibid., 4.

35. Ibid., 10–11.

36. Marvin Miles, "Four-Man Lab to Orbit in Space Sought: 4-MAN SPACE LAB," *Los Angeles Times*, April 24, 1963.

37. For Hilton's full speech, see Barron Hilton, "Hotels in Space," address given at the American Astronautical Society, 1967, http://www.spacefuture.com/archive/hotels_in_space.shtml (accessed July 10, 2017).

38. George E. Mueller, "In the Next Decade: A Lunar Base, Space Laboratories and a Shuttle Service," *New York Times*, July 21, 1969.

39. NASA was famously issued a fine for littering by the local Western Australian council, see: Emma Wynne, "When Skylab Fell to Earth," ABC Goldfields WA–Australian Broadcasting Corporation, July 09, 2009, http://www.abc.net.au/local/photos/2009/07/09/2621733.htm (accessed August 6, 2017).

40. Michael Hiltzik, "The Day When Three NASA Astronauts Staged a Strike in Space," *Los Angeles Times*, December 28, 2015, http://www.latimes.com/business/hiltzik/la-fi-mh-that-day -three-nasa-astronauts-20151228-column.html (accessed July 31, 2017).

41. Luke Harding, "Mars Mission in a Moscow Hangar Is No Joke, Say Astronauts," *The Guardian*, June 4, 2010, https://www.theguardian.com/science/2010/jun/03/mock-mission-mars-moscow -hangar (accessed July 31, 2017).

42. Ian Sample, "Fake Mission to Mars Leaves Astronauts Spaced Out," *The Guardian*, January 8, 2013, https://www.theguardian.com/science/2013/jan/07/fake-mission-mars-astronauts-spaced -out (accessed July 31, 2017).

43. The caption on the ESA Image database notes "the weightlessness wasn't simulated—except for fun by photo manipulation on 1 April 2011. This photo was sent by Diego Urbina via Twitter: 'Finally fixed the anti-gravity generation device!'"; see "Weightless on 1 April," European Space Agency, http://www.esa.int/spaceimages/Images/2011/04/Weightless_on_1_April (accessed December 20, 2017).

44. Latour and Woolgar, *Laboratory Life*, 17–18.

45. Papers of John F. Kennedy, Presidential Papers, President's Office Files, Speech Files, Address at Rice University, Houston, Texas, September 12, 1962, 3, https://www.jfklibrary.org/Asset -Viewer/Archives/JFKPOF-040-001.aspx (accessed July 10, 2017).

46. Joel Achenbach, "Trump Wants NASA to Send Humans to Mars Pronto—by His Second Term 'at Worst,'" *Washington Post*, April 24, 2017, https://www.washingtonpost.com/news/speaking -of-science/wp/2017/04/24/trump-wants-nasa-to-send-humans-to-mars-pronto-by-his-second -term-at-worst/ (accessed August 1, 2017).

47. Edwin (Buzz) Aldrin Jr., with Ken Abraham, *Magnificent Desolation: The Long Journey Home from the Moon* (New York: Crown Archetype, 2009).

48. Edward Everett Hale, "Life in the Brick Moon," *The Atlantic Monthly* 25, no. 148 (February 1870): 215–223 and 221; Edward Everett Hale, "The Brick Moon, III," *The Atlantic Monthly* 24, no. 145 (December 1869): 685.

10 The Urbane Laboratory: Applied Sciences New York

Russell Hughes

The socialization of scientists at all scales, from laboratory to park to precinct, is considered essential to provoking the chance fortuitous encounters that purportedly yield higher rates of scientific discovery and production. While strategies of spatial connectivity in scientific facilities have been attempted for well over half a century, today we see social and cultural infrastructures increasingly deployed to meet these ends. With the trend toward imagining entire cities as vast "urban laboratories" cosmopolitan or "global" cities find themselves ready-made in this regard, and thus in a distinctly advantageous position.

New York's quest to become "the new technology capital of the world" sees it invest in an elaborate scientific infrastructure that draws heavily on its considerable cultural resources.[1] Conflating the appurtenances of the city's finance, media, advertising, fashion, art, and entertainment "worlds" with those of a Silicon Valley start-up community, Applied Sciences New York (ASNY) constructs an alluring new scientific imaginary by which to attract the critical resources of talent and capital essential to "innovation." Anticipating every field of science to evolve until it becomes computer science, this boutique, bespoke brand of "digital urbanism" is designed as much for the histrionics or business of selling science as the doing of it at the laboratory bench. It conjures a speculative scientific milieu, one in which posture and projection are privileged over empirical merit.

The Urban Laboratory

The coming of the "urban century" anticipates nearly three billion additional people, mostly in developing countries, will enter cities over the coming thirty years.[2] While the prize economic opportunity in these emerging markets will be the combined purchasing power of these people entering the "consumer class," in the short to medium term it is the growth of the cities within which these consumer markets will boom that promise the most bountiful returns for investors.[3] The haste of urban development, and the unprecedented experiment in mass-scale conurbanism they produce, what the *Wall Street Journal* describes as "a threshold moment for humankind," questions the capacity of cities, already under extreme duress, to successfully accommodate this influx.[4] At this soteriological juncture, the commercial rhetoric of "smart cities" has broadened to encompass the experimental, "superorganismic" qualities of cities as "urban laboratories." Here, the developers of civic software promise not only to optimize and maximize the infrastructural performance of cities, but also to granularly gauge the health, interactions, opinions, activities, habits, swarms, moods, and *metabolisms* of their multifarious populations. Giving credence to the notion of the city as digitally "sense-able" or "sentient," this new species of digital urbanism produces a novel kind of "spatialized intelligence" from which a suite of innovative, smart civic commercial applications and opportunities ensue.[5]

With the city as the site, source, and now experimental *subject* of the scientific endeavor, it is municipally mandated that cities compete to attract the critical resources of capital and talent upon which twenty-first-century innovation and economic prosperity depend. Yet, what we are seeing in this competition is less a concentration on techno-scientific credentials, and more a focus on the capacity of cities to generate the *ambience* of innovation, as if atmosphere itself was an indispensable element of the scientific experiment.[6] Cities with vibrant bohemian cultures emanating metropolitan "buzz and fizz" are considered essential to the flourishing of the "creative class," for the socialization of scientists among their interdisciplinary selves, but also with members of the entrepreneurial sector whose venture capital and angel investment are fundamental to science's ongoing tenability.[7]

Of the thousands of cities across the world retrofitting and rebranding in this regard, the singularly most comprehensive, aggressive, and self-aggrandizing quest for technocratic civic transformation and global innovation domination thus far seen, is that of New York. Reminiscent of the way in which New York systematically overtook Paris as center of the modern art world in the aftermath of World War II, the smart civic branding at the heart of ASNY is similarly unique in its presentation of the next generation of "innovation" as one produced, and consumed, through a suite of elitist, exclusive "lifestyle" choices.[8] This chapter will analyze two major techno-scientific developments related to ASNY, Cornell Tech, and Hudson Yards, to demonstrate the ways in which the lifestyle qualities they express power and amplify the complex risk management techniques that underwrite the metroeconomics of smart civic innovation.[9]

"The New Technology Capital of the World"

In 2002, in the aftermath of the 1999 dot-com collapse and the 2001 terrorist attacks, New York's wealthiest citizen, Michael Bloomberg, was elected mayor on the basis that he could leverage his entrepreneurial talents and industry connections to reestablish the city as an economic world leader. One of Bloomberg's key aims was to diversify New York's economy to reduce its reliance on the vulnerable financial sector. Inspired by the success of homegrown tech companies Gilt Groupe, Etsy, Tumblr, and Foursquare, and the blossoming of high-technology startup districts Silicon Alley and DUMBO, Bloomberg envisioned New York as the place to merge finance, advertising, and media with emerging mobile platforms. Following extensive consultation, New York City Hall learned that a critical mass of digital engineering talent was needed to capitalize on this as yet underexploited tech niche. Though home to several high-ranking engineering schools, New York did not produce enough engineering graduates to compete with successful innovation districts like Silicon Valley or Boston's Route 128, nor did it have the necessary industries and associated incentives to keep them post graduation. To this end, the Bloomberg administration drafted an audacious plan to not merely make New York competitive with other high-tech centers but to eclipse them. Speaking to this scheme, U.S. Senator Charles E. Schumer declared—in the customary local vernacular—"look out Silicon Valley, look out Boston, New York will be second to none."[10]

On December 16, 2010, Bloomberg announced the Applied Sciences New York competition. Designed specifically to "increase the probability that the next high-growth company, a Google, Amazon, or Facebook, will emerge in New York City" it sought expressions of interest from academic institutions and/or joint consortiums to create a state-of-the-art applied science campus.[11] The competition stressed the importance of articulating links with corporate partners who would be "co-located" on site, "strongly" encouraging "proposals that also include space for related commercial activity such as business incubators, corporate research and development facilities, and spin out companies."[12] As incentive, City Hall would provide acreage on a number of prime development sites in the city, and up to USD 100 million in capital infrastructure expenditure and/or city-backed loans.[13] A little over one year later Bloomberg declared the winner: the joint proposal between Cornell University and the Technion Israel Institute of Technology to build a USD 2 billion, 2.2 million-square-foot (204,000 square meter) campus on Roosevelt Island.[14] Such was the strength of unsuccessful proposals that eighteen months later Bloomberg announced a second stage of ASNY, a Center for Urban Science and Progress (CUSP) led by New York University to be built in downtown Brooklyn.[15] This was followed by a third stage, the Columbia Institute for

Data Sciences and Engineering in Uptown Manhattan, and lastly the Carnegie Mellon/ Steiner Studios Digital Media Program to be built in the former Brooklyn Navy Yard.[16] Collectively the four ASNY projects are expected to generate over USD 33.2 billion in nominal economic activity, more than 48,000 permanent and construction jobs, and approximately 1,000 spinoff companies by 2046.[17]

Critical to this innovative endeavor is a Memorandum of Understanding between City Hall, ASNY's CUSP, and their corporate partners that center on the concept of New York as a "living laboratory."[18] With financial and mentoring support from technology companies IBM, Cisco, Con Edison, National Grid, Siemens, Xerox, AECOM, Arup, and IDEO, CUSP and its corporate partners will access New York's raw civic data to research and develop technologies that address the critical challenges and emerging growth opportunities in the provision of civic infrastructure, technology integration, energy efficiency, transportation, congestion, public safety, and public health.[19] City Hall's brokering of complementary ventures, such as CUSP's partnership with the Hudson Yards redevelopment, to be outlined shortly, further buttress ASNY's stake in the emerging field of "urban informatics." ASNY's state-academy-industry partnership seeks to corner this market in its infancy, cultivating a winner-takes-all "first mover" advantage by swiftly transferring local R&D into lucrative exportable products to aspirational urban laboratory markets across the developed and developing world.[20]

Despite Harvard Professor of Architecture Antoine Picon declaring that smart digital technologies "have not modified the physical structure of the city," two key ASNY-related projects under construction, Cornell Tech and Hudson Yards, strongly suggest otherwise.[21] ASNY's architectural and urban plan does not merely guide the hands-on "social" production of digital urbanism, for at this early stage these buildings also operate as the primary interfacial media and vehicle for the formal expression of the project's ideas and ideals. Concomitantly the projects function as engines of attraction to the venture capital and angel investment so critical to innovation's ostensible *ex-nihilo*. An examination of their sites yields much in the way of understanding ASNY's brand of twenty-first century techno-science as an ensemble production, and in particular, the importance of the projection of lifestyle as a critical scientific actant within it.

Cornell Tech

Situated prominently in the sight lines of the altitudinous boardrooms and corner offices of Midtown Manhattan, and the exclusive residences that line the East River, the area of Roosevelt Island south of the Queensboro Bridge is a premium location by which to visually showcase the source of New York's future economic fortunes. This

Figure 10.1
Cornell Tech, Stage One. Photograph by Philip LoNigro, 2017.

is home to ASNY's flagship campus, Cornell Tech, the first stage of which opened in 2017. No less impressive are the firms appointed to design this "iconic" site and its various programs from scratch: Skidmore, Owings and Merrill (SOM) are responsible for the campus master plan; Morphosis Architects, Weiss-Manfredi, Snøhetta, and Handel Architects for the campus buildings; and James Corner's Field Operations for the campus landscape.

The first thing to note in the site design is the naturally elevated circulation spine running through the center of the campus north to south. Situated nineteen feet (almost six meters) above sea level, the 'Tech Walk' follows Roosevelt Island's natural ridge. It seamlessly integrates landscape and built form, with pedestrian paths sloping right "up to the front doors and into the buildings."[22] Providing a sense of connection to all the elements of the site, the ridge additionally serves as a natural safeguard against a five-hundred-year flood event.[23] These ecological sensitivities foreground the more overt desire, as stipulated in the original ASNY Expression of Interest, for the campus to become the largest "net-zero" energy project ever built in the eastern United States.[24] To meet this goal, Cornell Tech is designed around a massive solar array (the "lilypad") and 400 geothermal wells that together generate enough power to supply all of the campus energy needs on site. These technologies concurrently function as a core feature of the wholly in situ living laboratory of sustainability research to be conducted in the Built Environment hub of Cornell Tech's academic curriculum. A 350-apartment residential tower, The House, designed by Handel Architects, socially engineers these green aspirations to an extreme degree, with residents required to actively monitor and minimize their individual carbon footprints.[25]

These impressive ecological credentials are nonetheless epiphenomenal to Cornell Tech's true test of sustainability, the generation of innovative energy between students, faculty, and corporate partners co-located on site. Thom Mayne of Morphosis Architects has the task of fulfilling this program in the first academic building, the Bloomberg Center. Mayne is a pioneer of the parametric approach to architectural design, and has completed numerous high-end educational facilities.[26] Yet despite this form in the field, he contends that a campus building such as that required for Cornell Tech has "no modern prototype. You have to have a completely different model which has to do with transparency ... exposing social connectivity and breaking down the Balkanization that happens departmentally."[27] Mayne's design resists the urge to predetermine the building's program around any particular notion or ideal of "innovation" lest its potential be limited. Mayne describes the "dialogue" of the building as articulating a complete "lack of privacy" and a "radical promotion of transparency," stating "it will be clear to someone visiting that this is not a typical academic setting."[28] This commitment to "openness" extends to the alignment of the building's atriums with those of Manhattan's grid across the East River, the interior stairwells at the western entrance "spill[ing] out" into "the canyon of 57th street." "You are," says Mayne, "literally a part of Manhattan."[29]

The open leitmotiv extends to the design of the campus's three departmental hubs that will be distinguished not by the customary separate building, wing, or even floor, but by modestly sized, permeable "hub lounges." For more focused parties, zones with workstations, "huddle rooms," "swing spaces," and "collab rooms" will be on hand. Once promising discovery reaches the pointy end of transfer negotiation, "entrepreneurial patios" will present the requisite breathing space to nut out the all-important distribution of intellectual property. This last activity is something the co-location or Bridge building by Marion Weiss and Michael Manfredi is designed specifically to foster.[30] It is overshadowed by the imposition of the lilypad in what one commentator describes as being in the "headlock of an older, stronger brother."[31] This doesn't dampen Marion Weiss's excitement, however, at being offered an "opportunity unlike any other we can think of."[32] Charged with the task of cleaving the triple-helix partners of academy, civic government, and co-location corporate sponsors together, Weiss-Manfredi see the Bridge as an opportunity to experiment *within* an experiment: "This campus

Figure 10.2
The Emma and Georgina Bloomberg Center (Morphosis Architects) and the Bridge Building (Weiss-Manfredi Architects). Photograph by Philip LoNigro, 2017.

here is an invention already, and this particular building, a co-location building, is even more of an invention. Here's a case where there's a completely catalytic concentration of academia, industry, and innovation in one place."[33] Weiss's partner, Michael Manfredi, similarly contemplates, "how can we think of creating a kind of architectural, in a way 'hothouse' that can make different kinds of people come together?"[34] They contend that carving the building in two—"and from that center an ability spilt down, and up, so that the disconnection that one could experience in a core building is now transformed into a catalytic connection between the landscape, between the floors, and between the tenants, both academic and entrepreneurial"—will render the "disassociation normally felt in [an] academic silo ... erased."[35]

This architectural erasure is a synecdoche for ASNY's broader desire to eliminate the barrier between higher education and industry. Inspired by the tech ecosystem that blossomed around Stanford University, yet conceived along the lines of tech incubators such as Y Combinator, Techstars, and General Assembly, Cornell Tech's co-location corporate partners pay in the order of tens of millions of dollars to not only gain access to experimental research, but also literally prescribe industry specific problems for graduate engineering students to solve.[36] Though the new institute offers unique "dual"

Master of Applied Science degrees, that is, one from Cornell *and* one from the Technion, the rites and rituals of the traditional university, and the production of first-class graduates for industry, are perfunctory and ancillary to the campus's primary intention: the realization of student potential in the present tense. This shift in the nature of intellectual property production, the very thing that most high-technology companies stake their future fortunes on, is best explained by the wunderkind origins of many iconic tech enterprises. As per ASNY's stated intention to produce "the next Google, Amazon or Facebook," according to the popular mythology, high-tech success begins with one or two young, western, white, middle-class men. Immersed in a mélange of hardware and software, and in possession of a prodigious, precocious, and prescient technological insight, they assemble all the elements at their disposal to produce a breakthrough product that becomes a socio-cultural "game changer," commercial "cash cow," and foundation by which to launch a litany of other "world changing" ventures. Popular culture—indeed the very network of digital infrastructure upon which popular culture is disseminated—is replete with examples. Contrary to the fragmentation of identity popularized at the dawning of the age of the ubiquitous screen by Sherry Turkle, echoing Frederic Jameson's earlier "schizophrenic" postmodernism, Picon contends that the era of mass digitization, and its superstar CEOs, enable the reconstruction of the sovereign "heroic figure."[37] He observes that the ebullient nature of smart city discourses tend to have a "self-fulfilling" character in this regard, generating "the conditions that make them feasible, in the same way as some political or economic forecasts influence voting dynamics or market behavior by causing them to lean in the direction that makes them possible."[38] The challenge, and opportunity, for ASNY, is how to transfer the hero mythology from the (Stanford) dropout hacking away in his parent's Northern Californian garage, to a willingly enrolled ASNY student-resident-citizen.

A closer examination of the way the coordinates of "openness" and "erasure" pass beyond Cornell Tech's epidermal walls, to their exterior expression in landscape, explains ASNY's strategy for achieving this aim. SOM's pedestrian-centric master plan links the public spaces on campus to those on the island, in particular the Franklin D. Roosevelt Four Freedoms Park at its most southern tip. New York's Director of Capital Projects and Planning Andrew Winters says the goal of the master plan is for "outside" people to "feel like they're invited."[39] Despite evidence suggesting this may be a difficult ideal to execute, Colin Koop of SOM elaborates on Winters's "open" intentions.[40] Koop refers to the sinewy crossings on the master plan as "paths of desire" people will naturally traverse, privileging positions that offer the best vantage to take in the "cinematic" views of Manhattan's architectural marvels.[41] Reminiscent of similar strategies used across town in the design of the celebrated Chelsea High Line park, Cornell Tech

Figure 10.3
The Emma and Georgina Bloomberg Center at Cornell Tech. Photograph by Sperli, Wikimedia Commons.

uses the same approach—indeed the very same landscape architect of the High Line, James Corner of Field Operations—to reproduce this cinematographic effect. Mimetic of scenes lifted direct from Woody Allen's iconic 1979 film *Manhattan*, Cornell Tech's landscape exploits the site's panoramic vistas, providing an enticing new "must-see" designer place for souvenir hungry *flâneurs* to flock—selfie poles in hand—en masse.

Yet attracting people *into* the orbit of Cornell Tech is, however, only the first stage of the more ambitious landscape plan to centripetally propel the campus *out* into the wider city. Corner calls his design for the "open space ground-plane" of Cornell Tech's landscape a "tissue that's going to hold it all together, so that it's not just an ensemble of really great buildings, but something held by the *matrix*."[42] Corner's reference to this mathematical principle, where all scales equal one another in order to multiply their effect, alludes also to "the cultural, social, and political environment in which something develops."[43] Here, New York City Hall plans for Cornell Tech go beyond educating the next generation of digital engineering talent; by ensuring access to affordable real estate and infrastructural amenities, it also hopes to facilitate their entrepreneurial expansion, thereby keeping that talent local once ripe. Former Cornell Tech Vice President Cathy Dove envisions a "spillover" of innovation extending directly from the campus, "flowing both west and east" along Roosevelt Island's main transport routes.[44] Long Island City's relatively affordable, burgeoning creative communities gathering around MoMA's PS1 and Silvercup Studios further south are critical destinations for this expansion, as are select districts in the outer boroughs of the Bronx and Harlem where tech incubators have already been kickstarted with funding from the Bloomberg administration.[45] To incentivize this viral spread, Dove says Cornell Tech is establishing strategically placed "landing pads" throughout the city that will be the go-to destinations for graduate success stories in swift need of expansion.[46]

Whether or not ASNY's industry-based curricula dispersed throughout urban sentinels will be enough to lure the next generation of wunderkind digital innovators from Silicon Valley and Boston remains to be seen. As Greg Pass, former chief technology officer at Twitter, and now Cornell Tech's "Founding Entrepreneurial Officer" concedes, the new institute is still "at a stage where we're challenging all assumptions and figuring out what the DNA of this new type of school should be."[47] A closer forensic inspection of the project's architecture and urban plan reveals however, and in no uncertain terms, who the progenitor of ASNY's multiplicative source code is.

On SOM's master plan, "activation" lines are sketched for the intended indoor-outdoor flow from ground to building. The paths merge into the mouth of Mayne's double-height skylit café situated adjacent to a lecture hall and exhibit gallery. "You've been to the Ace Hotel?" Mayne asks in an interview. "Their lobby is what this space is."[48] Epitomized as the ideal place to do "casual business," the open plan of the Ace Hotel lobby, located on West 29th Street in Midtown Manhattan, has since been popularized in the spatial organization of many big digital companies. A well-known local derivative is the sixth-floor entry point at Bloomberg L.P., former New York Mayor Michael Bloomberg's privately owned financial software, data, and media company.[49] A mandatory elevator stop in his 731 Lexington Avenue skyscraper, Bloomberg L.P.'s sixth-floor lobby tactfully weaves professionalism with pleasure as an array of open expanses, free food points, and commingling areas make it a socially salubrious "gateway" experience to the other fifty-four floors of the building.

Cornell Tech's ostensible dedication to "erasure" and "openness" therefore should not be misinterpreted as "empty," for laced throughout the campus reproduction of iconic city sites, targeted curricula, and projected future alumni profile is the more visceral desire to reproduce, at scale, New York's most successful entrepreneurial formula. Bloomberg is an engineer by training whose company made its fortune selling innovative financial software, data, and media solutions to Wall Street. Bloomberg L.P. is the single most successful technology company ever started in New York, and Bloomberg is its wealthiest citizen, incidentally the sixth richest person in the United States, the eighth richest in the world.[50] In his decade-long tenure as mayor, Bloomberg's retrofit of City Hall's buildings transformed his own administration space into something resembling an open-plan trading floor. Affectionately coined "Mike's bullpen," it spoke eloquently of Bloomberg's move into, and viral transformation of, state bureaucracy. Given ASNY is his brainchild, the apotheosis of his mayoralty, and likely recipient of much of the USD 49 billion fortune the seventy-five year-old intends to give away in his lifetime, Bloomberg's technocracy now extends through New York's revamped academe with the intention of virally disseminating throughout the entire city economy.[51] As with the naming rights for Cornell Tech's first academic building that Bloomberg bought (in his daughters' names, not his) for a USD 100 million donation, ASNY more broadly wears his noblesse oblige signature stamp, the project nothing if not a citywide program of social engineering more or less cloned in his image.

Intellectual Property: The Hudson Yards Redevelopment

Another project initiated during Bloomberg's mayoral tenure speaks even more voluminously of these ambitions. The Hudson Yards redevelopment, at a cost of over USD 20 billion, is America's "biggest ever real estate project."[52] Hudson Yards is being built on a platform covering six city-sized blocks of existing railway yards on the Westside of Midtown Manhattan. On completion in 2027 it will provide over eighteen million square feet (five million square meters) of mixed residential, office, and retail space.[53] Chris Smith describes it as "the quintessential expression of Bloombergism: big, expensive, highly manicured, with a fine-arts sheen and serious insider connections."[54] As striking as its extraordinary capaciousness is the heraldic claim that Hudson Yards will house the world's first-ever "quantified community."[55] In a high-profile coup for ASNY, its Center for Urban Science and Progress (CUSP) campus has established a partnership with Hudson Yards to access all data from its embedded sensors, biometric recognition software, and personal communication devices operated within it. Using this information, CUSP will monitor and analyze pedestrian flows, vehicular traffic, air

Figure 10.4
Hudson Yards Redevelopment. Courtesy of Kohn, Pedersen & Fox Associates, 2017.

quality, energy use, waste disposal, and the health and activity of participating residents, employees and visitors.[56] CUSP envisions this real-time biometrically contained urban data laboratory as potentiate for intelligently understanding and intervening in the rest of the New York metropolitan area. It offers a prophetic glimpse into the future of civic administration in not only New York, but, in accordance with ASNY's ne plus ultra export ambitions, all other emulative urban laboratories beyond.[57]

On one level, civic renewal ventures such as Hudson Yards, Cornell Tech, and ASNY's three other new campuses constitute the latest iteration of City Hall's mission to curate a bouquet of nouveau New York "experiences" for residents, tourists, and the film industry alike. Part of Bloomberg's original 2002 economic recovery plan, the renewal of underutilized urban assets is part of a much broader project of civic revitalization underway since the 1960s. Arguably as much about "value-adding" beautification and gentrification as they are, in this case, about scientific education, research, and innovation, raising New York's property values socially engineers, in a process of stigmergic autophagocytosis, the city's citizenry toward higher socioeconomic brackets, activating a series of economic levers benefitting City Hall's bottom line. Furthermore, ASNY and related projects' triple-helicoid *financial* engineering generates transactional activity stimulating its own kind of pecuniary bull market, a game New York more customarily excels in.[58]

Yet there is another dynamic at play in this complex economic equation, one that courses deeper than the characteristically obvious or cosmetic. ASNY and related ventures, in particular Hudson Yards, are symptomatic of what are defined as *paquebots urbains*.[59] Translated literally as "urban cruise ships" or "urban liners," these large-scale "hypertechnical, integrated and multifactional" developments with massive interiors possess "all the essentials of public life in installation."[60] They function as autonomous worlds seemingly independent of the city in which they are situated. Dominique Lorraine attributes the emergence of these elephantine architectural-urban enclaves to the financialization of the late twentieth and early twenty-first centuries.[61] Between 1993 and 2010, over half of the increase in national income in the United States went to the top 1 percent of households.[62] The years between 2009 and 2012 prove this to be an exponentially accruing trend, with 95 percent of total income gains ending up in those very same hands.[63] For New York's seventy-nine billionaires and nearly four hundred thousand millionaires, a city in neck-and-neck competition with Beijing and London for "capital of the world" by way of having the most of it, the construction of ever more reified *paquebots urbains* is critical to attracting ever more capital by first and foremost securing or "risk-managing" the interests of that capital already embedded in place.[64] The importance of data security to this equation cannot be overstated. With automatic trading algorithms responsible for nearly 70 percent of all financial trades in the United States, most of which occur in mere fractions of a second, such is the criticality of trading speed that a new fiber-optic cable linking Chicago to New York was built in 2009 for the purpose of offering three to four thousandths of a second faster delivery than existing communications routes.[65] A new cable between London and New York opened in 2015 at a cost of USD 300 million providing "unprecedented" transmission speeds of "sub-58.95" milliseconds.[66] Be it Hudson Yards with its looped fiber network and on-site electric microgrid, among other utilities, housed forty feet above sea level, or Cornell Tech's energy self-sufficiency and elevation safeguarding it against power outages and/or a 500-year flood, NYC's new moon bases on either side of midtown Manhattan offer more than mere physical refuge amid Hurricane Sandy-like disasters.[67] They also register as panoplies of *financial* fortification, guaranteeing the bandwidth to future proof capital—that cannot afford to be out of the financial game for a femtosecond—from the inevitability of rising sea levels and/or increased meteorological activity.[68]

In this last respect, ASNY's ostensible commitment to "innovation" translates more as risk aversion. It renders the city not so much "smart" as paranoid. This contradictory logic reaches its apogee in the way "intelligence," the purported currency of the smart city, is not invested in its creative populace, but rather is *outsourced* to autonomous third parties.

Acceleration and Enclosure

ASNY's urban plan is generated using the now-ubiquitous techniques of digital para-metricism. Benjamin Bratton is suspicious of the upscaling of this design approach, wherein the "smart city" is determined by "dense material and logistical equation[s]" that operate like "spreadsheets waiting for the right formulas," their primary design interests "the flows of algorithmic capital and its spastic valuations of land, energy, information, and human capital."[69] Bratton's argument is reminiscent of the aforementioned "a-signifying semiotics"—mentioned in the Introduction to this volume—that Maurizio Lazzarato argues power and amplify the "silent mechanisms" of "mathematics, stock quotes [and] business."[70] Considered in the context of the complexities pertaining to planetary scale computation today, Bratton discerns a significantly more enveloping sign-system; an "accidental megastructure" he identifies as *The Stack.* According to his *Stack* schema, the smart city-cum-urban laboratory is not an end in itself, but rather is one of six "layers." Listed from bottom to top as *Earth, Cloud, City, Address, Interface,* and *User* layers, what we see with the advent of the smart city is the "telescoping [of] descriptive and prescriptive algorithms set in motion in the *Cloud* layer ... building their local franchises into urban (or 'urbanish') investments."[71] Here the clash between atoms and bits, represented in the manic evolution of glass optic fiber granularly mashed into the concrete and steel of the civic envelope, increasingly forms a series of "interfacial partition[s]" that open or close "urban spaces to different *Users* in different ways," their primary role "sorting ... *Users* in transit."[72]

Building on this "sorting" thesis, "by way of analogy and allegory," ASNY can be seen as a project that algorithmically transforms New York into its own kind of "search engine," the *City* coded to conjoin "innovative" components (the raw elements of capital, talent, and other techno-scientific resources) complementarily together. A contemporaneous rewriting of the fable of the philosopher's stone, ASNY's organizational and urban choreography torrents the critical ingredients it needs to alchemize more than the sum of its parts breakthrough "innovation" for the purpose of yielding, it hopes, high-growth, high-tech corporate success.

The search engine analogy achieves startling clarity when considered in the context of one last, but crucial element in New York's digital urban plan—the 2.1-kilometer western extension of the number 7 line of the city's subway system. As previously mentioned, for *Users* heading east from Cornell Tech, the subway takes just two minutes to get to the "landing pad" area of Western Queens. From there another twenty-five minutes south reaches ASNY's CUSP and Carnegie Mellon/Steiner Studio campuses, and the booming tech districts of Brooklyn and DUMBO. Five minutes back across the East

River is Wall Street, another ten to fifteen minutes north delivering *Users* to the startup hotbeds of "Silicon Alley" Union Square and/or the Chelsea-Meatpacking tech district. Conversely, going west from Roosevelt Island, it takes just nineteen minutes, by connection, before arriving at the destination of the number 7 extension—at a cost of USD 2.4 billion to the New York taxpayer—the new 34th Street station at Hudson Yards.[73] Forming a C-shaped arc in reverse, these routes seamlessly converge at either end of the Chelsea High Line. Leaving the northern end of the High Line by foot, a delightfully "cinematic" twenty-five-minute stroll south above the traffic connects this innovative circle that ends, only to begin again, on the border of the Chelsea-Meatpacking tech area, home to Google, the largest single-owner technology space in the city.

Nowhere is ASNY's connection to New York's corporate sector any more pronounced, either geographically or organizationally, than it is with Google. The archetype and exemplar for the kind of high-growth company ASNY hopes to produce, Google's gigantic Chelsea offices, the former Port Authority of New York building, has been home to Cornell Tech's pilot program since 2012 where it has run rent-free.[74] Furthermore, Cornell Tech has struck a deal with Google to outsource the supply of its all-important digital infrastructure to the company's prodigiously evolving cloud platform offsite, also for free.[75] Testament to Google's increasing presence in the city, and intimacy with New York City Hall, is Sidewalk Labs (a subsidiary of "Alphabet," Google's recently anointed "parent" company) underwriting of the new LinkNYC project. Converting hundreds of New York public payphones into state-of-the-art Internet terminals and Wi-Fi connectors (that additionally possess the capability to function as data exchanges for autonomous vehicles, public transit, and other nascent urban systems) free to the public, cost Sidewalk Labs in the vicinity of a half billion dollars.[76] The company will initially monetize LinkNYC through the customary channels of on-street advertising and the selling of data New York *Users* provide, the latter revenue stream prompting Nick Pinto to caution, "If you're not paying for the product, you *are* the product."[77] Pinto suggests LinkNYC is a "radical step even for Google. It is an effort to establish a permanent presence across our city, block by block, and to extend its online model to the physical landscape we humans occupy on a daily basis."[78] Despite City Hall's policy, enshrined in law, of "Open Data NYC," and Thom Mayne's best intentions, Alpha Google's objectives in and for New York remain, at this early stage, anything but "radically transparent."[79] Nevertheless, what this organizational husbandry does reveal is how New York has, in accordance with Bratton's "sorting" thesis, metaphorically *and* literally become a three-dimensional search engine that "privileges" certain endeavors over others.[80] ASNY's urban plan, scripted both *like* and *by* page rank algorithms, is geared toward the "innovative" transformation of the *City's* raw elements and refinement of its flows. This powering and amplifying of *Stack*-like vectors throughout New York similarly transforms and refines individual scientists, and in particular, the *style* of science they come to produce.

Homo Stylus

Though this volume declares lifestyle to be, in the Latourian sense, a critical "actant" in the production of science, ASNY's organizational, architectural, and urban plan does not make explicit, or easily communicate at this early stage, why this is so. In the context of seeking to become the new technology capital of the world, a reflex reading finds that New York's abundance of metropolitan buzz and fizz easily trumps Silicon Valley's innovation-rich, but culturally starved, suburb-banality. Despite not having California's pool of talent or abundance of speculative venture capital, what New York's tech ecosystem does have in spades are its enviable *urbane* qualities. New York is the undisputed global heavyweight of "Coolhunting" in this regard, made ever more so in the computational context by the city's colonization of digital urbanism's popular imaginary in television serials such as *Person of Interest, Mr. Robot,* and the like.[81] Yet there are more compelling reasons why lifestyle is so important to the success of this smart city. Aesthetic and sensual concerns have become more precious proportionate to the deranged virtualization threatening them. Echoing the earlier return to place in urbanism, a concomitantly reactive return to tangibility has also ensued, something *Smart Cities: A Spatialised Intelligence* author, Antoine Picon, argues is led by the creative class who demand sensual gratification in an increasingly ephemeralized civic sense-scape:

> Tactility in particular constitutes a rapidly rising dimension in contemporary culture [evidenced] by the new place accorded to it by architecture through what tends to be described as the "return" of ornament. … Far from being cut off from sensation by the digital revolution [the individual] reveals him- or herself to be hyper-receptive to all types of sensory stimuli. … Scientists, businesspeople or designers, are striving for a rich and varied urban environment that engages all five senses. Art galleries, performance venues, gastronomic restaurants and fashion boutiques all pertain to the ecology that the knowledge economy requires, just as much as sensors, fibre optics and ubiquitous computing do.[82]

ASNY's architectural and urban appropriation of New York's preexisting cultural infrastructure sets the gold standard for sybaritic science, the city's dynamic cosmopolitanism befitting of, and thus attractive to, the new species of digital capital and talent ASNY needs to fulfill its technocratic ambitions. As evidenced by Cornell Tech's pilot "Runway Startup Postdoctoral Model," "Student Spotlight" and "Startup Awards,"

Figure 10.5
Link NYC Booth, West 14th Street. Photograph by the author, 2017.

ASNY already practices a demagogic catwalk-science where one-of-a-kind innovation struts its panache to the city-centric anthems of Frank Sinatra, Jay Z featuring Alicia Keys, or Taylor Swift; the legacy of the Fricks, Astors, Vanderbilts, and Carnegies never far from view.

Yet, compelling as these diagnoses are, *still* they fail to capture the quintessential reason why the exhibition of an urbane lifestyle is so critical to the success of today's scientists and, by proxy, that of the urban laboratories in which they ply their trade. As evidenced by the rate at which Google and other big digital companies are entering the scientific market, if computational power, big data storage capacity, and artificially intelligent algorithmic panopticons are key to techno-scientific breakthrough, ASNY's entire program correspondingly banks on the assumption that every field of science evolves until it becomes computer science. As with most other organizations, be they private companies, NGOs, not-for-profits, or, now, state administrations, ASNY partners with a cloud provider to not only thrive, but survive, in a new geopolitical reality where cloud platforms challenge the traditional sovereignty of the Westphalian nation state.[83] The incremental outsourcing of core organizational functions to big digital companies is similarly reflected in the outsourcing of traditionally skilled occupations—from finance to design, and now science—to third-party algorithmic governance. This outsourcing is key to explaining the shift toward the expression of a civic lifestyle as part of one's vocation. In the case of science, automation does not necessarily mean the de-skilling of techno-scientific labor, but rather a shift in its mode of production. To gain funding to pursue promising research, scientists must increasingly fulfill the roles of strategist, lobbyist, entrepreneur, and businessperson. They must not only coordinate diverse teams of scientific expertise, but also engage with lawyers, economists, bioinformaticians, big data IT specialists, and a suite of other actors and stakeholders necessary to scientific production. Interdisciplinary performers par excellence, scientists, techno-scientists, or more broadly "innovators" as we now must call them, similarly have to negotiate the financialization of science, where ten- to fifteen-year research projects become traded like any other futures commodity or hedge fund, subject to renewal or abandonment on a quarterly basis.

This shift is evidenced in the diminishment of actual laboratory space in contemporary research buildings, today characterized by kernel laboratories dwarfed by administrative and communication facilities. The proportion of research facilities dedicated to wet laboratories will continue to shrink as stand-alone robots, capable of performing the routine work of 10,000 bioscientists, are developed and ultimately deployed at scale.[84] Adding insult to injury, the dawning of the big data age furthermore reduces the need for a scientist's *intellectual* expertise. As signaled in the groundbreaking 2009 journal article "The Unreasonable Effectiveness of Data" (incidentally authored by

three of Google's top-ranking artificial intelligence engineers), the millennial-long scientific obsession with exactitude and precision is giving way to the ever-increasing
power of algorithmic probability.[85] As Viktor Mayer-Schönberger and Kenneth Cukier confirm, "Though it may seem counterintuitive first, treating data as something
imperfect and imprecise lets us make superior forecasts, and thus understand our world
better."[86] Algorithmic software can correlate noncausal relationships in huge data sets
from many different angles, enabling us to see links we never saw before, and grasp
complex technical, scientific, and social dynamics that otherwise escape the limits of
our organic cognitive comprehension. Thus, department stores can identify pregnancy
in customers before they even know it.[87] Banks can predict divorce.[88] And FICO's CEO
can claim he "knows what you're going to do tomorrow."[89] These noncausal analyses
aid our understanding not by asking "why" something might work, then proving it,
but rather simply asking "what" works; that is, what massive data sets coupled with
correlative, pattern-recognition algorithms tell us. In this context Chris Anderson proposes "the data deluge makes the scientific method obsolete."[90] He argues that companies like Google, having aggregated the Internet into a single database that acts as a
"massive corpus" and "laboratory of the human condition," bring us to a point where
the more we learn, "the further we find ourselves from a model to explain it."[91] Anderson contends that with huge amounts of data and new statistical tools, "correlation
supersedes causation, and science can advance even without coherent models, unified
theories, or really any mechanistic explanation at all."[92] As the petabyte age dawns, it
does away with the need for both a hypothesis and a scientist to generate it.

Though the tech sector is booming, with tech firms representing 21 percent of the
largest 500 American companies, they employ only 3 percent of its workforce.[93] Stag-
nant living standards, rising inequality, and what is called a growing "precariat" (precarious proletariat) in the "gig economy" are testament to the steady outsourcing of
techno-scientific labor and expertise to third-party algorithmic governance.[94] Struggling
to maintain vocational relevance, scientists-cum-innovators have to constantly reinvent themselves, thus increasingly "perform" science on celebratized, *urbane* "world"
stages, such dramaturgy critical to projecting the confidence necessary to building relationships with the capital and talent so vital to research in these aleatory scientific
times.[95]

Conclusion

Planetary-scale computation telescopes itself through the metrocononics of the smart
City into de novo civic scientists. No longer precise, certain, or secure, these scientists
can neither afford to be conservative, nor hesitate in what they can promise to leverage

Figure 10.6
Larry Page and Michael Bloomberg, Cornell Tech Pilot Announcement, Google's New York offices.
Courtesy of Newscom, 2012.

from the synthetic system/regime/machine/*Stack*/meta-assemblage (call it what you will). Using cloud systems to see what our organic, biological eyes alone cannot, having a foot in both Aristotelian *and* Platonic camps, the "method" of lead scientists today walks a slippery slope of ostentation and overcompensation in order to meet funding demands hitched to the exponential arc of innovation's ever-accruing hyperbole. Whether Bloomberg's innovation blitzkrieg of New York, both consumer *and* producer of this trend, will in Terman-esque fashion fulfill the projects' techno-imperialist charter, only time can tell.[96] The bigger question is whether ASNY's éminence grise anticipated the extent to which Google, doing a little leveraging of its own, would use its algorithmic altruism to Trojan horse ASNY and hijack (hack) New York along with it?

Though strong analogies can be drawn between the brash optimism of ASNY and the brazen confidence upon which New York's finance, media, advertising, fashion, art, and entertainment worlds are run, what is really "powered and amplified" through this very site-specific *City* layer appears more everyday a new kind of accident.[97] As pervasive, ambient artificial intelligences take hold of the global innovation sector, their deployment essential to overcoming (outsourcing) humanity's biggest challenges and threats, there is no telling what bottom-up, sideways, and diagonally "deep" deviations will emerge from this classic top-down megaventure, other than what history makes abundantly clear: that chimerically emerge they will! New York Science City will ultimately resemble neither a Bloomberg Terminal nor Google's all-consuming algorithmic business model, but rather the monstrous more than the sum of them both. Indeed, if most of the Internet's traffic is already generated by nonhuman *Users*, the real currencies coursing throughout this smart *Cloud City* are complex mutations whose ongoing evolution eludes reference to anything ever before seen.[98] Bratton cautiously gestures toward "fast Darwinism['s] … inhuman and inhumanist molecular form finding."[99] Kevin Kelly is a little less circumspect, his *Stack* a "Technium" he argues ("inevitably," and irreversibly) is just now awakening, "whispering to itself."[100]

Prior to Google's release in 1996, a "perplexed" Kelly asked Google cofounder, Larry Page, why he was building another search engine given "Alta Vista … seemed good enough." Page answered, "Oh, it's not to make a search engine, it's to make an AI."[101] Two decades on, the zeal with which Google has co-opted its latest acquisition is critical, it would appear, to the metaphorical *and* literal fulfillment of what one astute investor identifies as the company's overarching mission: to be the "Manhattan project for AI."[102]

Notes

1. This term "the new technology capital of the world" was used by a number of dignitaries at the announcement of Cornell Tech winning the Applied Sciences New York competition, most

notably Senator Charles E Schumer. See "Cornell Wins NYC Tech Campus Bid," *Cornell Chronicle*, December 19, 2011, http://news.cornell.edu/stories/2011/12/cornell-wins-nyc-tech-campus-bid (accessed March 11, 2014).

2. United Nations, "World Urbanization Prospects: The 2011 Revision," Department of Economic and Social Affairs Population Division, 2012, http://www.un.org/en/development/desa/population/publications/pdf/urbanization/WUP2011_Report.pdf (accessed November 13, 2013).

3. What is perhaps most extraordinary is that this growth miracle will occur in roughly 440 cities in developing nations, many of which are still today only subsistent rural outposts. See Richard Dobbs, James Manyika, and Jonathon Woetzel, *No Ordinary Disruption: The Four Global Forces Breaking All the Trends* (New York: Perseus, 2015), 125, Kindle edition.

4. Robert Lee Holtz, "As World Crowds in, Cities Become Digital Laboratories," *Wall Street Journal*, December 11, 2015, https://www.wsj.com/articles/as-world-crowds-in-cities-become-digital-laboratories-1449850244 (accessed December 11, 2015).

5. "Sentient city and sensory or even sensual city: the two perspectives are interrelated. The notion of the senseable city, to which the laboratory founded by Carlo Ratti at MIT refers, purposely plays on the confusion between these two possible interpretations of the urban realm." See Antoine Picon, *Smart Cities: A Spatialised Intelligence* (Chichester, UK: Wiley, 2015), 44. Also see MIT Senseable Cities Lab, http://senseable.mit.edu/ (accessed May 1, 2017).

6. Kai Wen Wong and Tim Bunnell, "'New Economy' Discourse and Spaces in Singapore: A Case Study of One-North," *Environment and Planning A* 38 (2006): 38–78.

7. Peter Hall, *Cities of Tomorrow* (Oxford: Blackwell, 2002), 408; Richard Florida, *The Rise of the Creative Class* (New York: Basic Books, 2002).

8. See Serge Guilbaut, *How New York Stole the Idea of Modern Art* (Chicago: University of Chicago Press, 1985).

9. "Metroeconomics" is a term from Benjamin Bratton, *The Stack: On Software and Sovereignty* (Cambridge MA: MIT Press, 2015), 159–160.

10. Office of the Mayor of the City of New York, "Cornell Wins NYC Tech Campus Bid," *Cornell Chronicle* press release, December 19, 2011.

11. "Applied Sciences NYC," New York City Economic Development Corporation (NYCEDC), September 12, 2016, https://www.nycedc.com/project/applied-sciences-nyc (accessed March 1, 2014).

12. "Mayor Bloomberg Announces Initiative to Develop a New Engineering and Applied Sciences Research Campus to Bolster City's Innovation Economy," NYCEDC, December 16, 2010, https://www.nycedc.com/press-release/mayor-bloomberg-announces-initiative-develop-new-engineering-and-applied-sciences (accessed March 11, 2014).

13. Ibid.

14. "Cornell Wins NYC Tech Campus Bid."

15. "Mayor Bloomberg, New York University President Sexton, and MTA Chairman Lhota Announce Historic Partnership to Create New Applied Sciences Center in Downtown Brooklyn," Office of the Mayor of New York City, April 23, 2012, http://www1.nyc.gov/office-of-the-mayor/news/147-12/mayor-bloomberg-new-york-university-president-sexton-mta-chairman-lhota-historic (accessed March 11, 2014).

16. "Mayor Bloomberg and Columbia University President Bollinger Announce Agreement to Create a New Institute for Data Sciences and Engineering," *New York City*, July 30, 2012, http://www1.nyc.gov/office-of-the-mayor/news/280-12/mayor-bloomberg-columbia-university-president-bollinger-agreement-create-new#/1 (accessed March 11, 2014).

17. "Mayor Bloomberg Announces Carnegie Mellon University Will Open Fourth New Applied Sciences Program in NYC," Office of the Mayor of New York City, November 20, 2013, http://www1.nyc.gov/office-of-the-mayor/news/376-13/mayor-bloomberg-carnegie-mellon-university-will-open-fourth-new-applied-sciences/-/0 (accessed March 11, 2014).

18. "Mayor Bloomberg, New York University President Sexton."

19. Barbara Murray, "NYU CUSP, Related & Oxford Team to Create First 'Quantified Community' in U.S. at Hudson Yards," *Commercial Property Executive*, April 15, 2014, https://www.cpexecutive.com/headlines/nyu-cusp-related-oxford-team-to-create-first-quantified-community-in-u-s-at-hudson-yards/1004094679.html (accessed November 11, 2015).

20. Ibid.

21. Picon, *Smart Cities*, 14.

22. Cornell University, "Cornell Tech: Designed for Impact," January 7, 2014, https://www.youtube.com/watch?v=PFRIKri9Y_c (accessed October 24, 2014).

23. Alexandra Lange, "Silicon Island," *The New Yorker*, October 15, 2012, https://www.newyorker.com/culture/culture-desk/silicon-island (accessed June 9, 2014).

24. "Mayor Bloomberg, Cornell President Skorton and Technion President Lavie Announce Historic Partnership to Build a New Applied Sciences Campus on Roosevelt Island," Technion Israel Institute of Technology, December 12, 2011, http://www.technion.ac.il/en/2011/12/mayor-bloomberg-cornell-president-skorton-and-technion-president-lavie-announce-historic-partnership-to-build-a-new-applied-sciences-campus-on-roosevelt-island/ (accessed March 17, 2014).

25. "The House at Cornell Tech," Handel Architects LLP, http://www.handelarchitects.com/projects/project-main/cornell-res-main.html (accessed April 14, 2017).

26. Mayne's Morphosis has designed a score of educational facilities, many of which have been built. The more recent of these include the A. Alfred Taubman Engineering, Architecture, and Life Sciences Complex at Lawrence Technological University, Southfield Michigan; Emerson College Los Angeles; and 41 Cooper Square, the Cooper Union for the Advancement of Science and Art, New York. See Educational works, Morphosis Architects, https://www.morphosis.com/architecture/type?q=educational (accessed May 2, 2017).

27. Robin Pogrebin, "CornellNYC Chooses Its Architect," the *New York Times*, May 8, 2012, http://www.nytimes.com/2012/05/09/arts/design/thom-mayne-of-morphosis-is-chosen-for -cornellnyc-tech.html (accessed March 15, 2014).

28. Cornell University, "Cornell Tech: Designed for Impact."

29. Ibid.

30. Avi Wolfman-Arendt, "Creating an Ever-Flexible Center for Tech Innovation," *New York Times*, August 10, 2014, https://www.nytimes.com/2014/08/11/education/creating-an-ever -flexible-center-for-tech-innovation.html (accessed November 3, 2015).

31. Lange, "Silicon Island."

32. Cornell University, "Cornell Tech: Designed for Impact."

33. Ibid.

34. Ibid.

35. Ibid.

36. "Two Sigma to Be First Tenant at The Bridge at Cornell Tech," *Cornell Chronicle*, Jan 24, 2017, http://news.cornell.edu/stories/2017/01/two-sigma-be-first-tenant-bridge-cornell-tech (accessed May 2, 2017). Verizon also are a co-location sponsor, though they do not inhabit the Bridge. See "Cornell Tech Announces $50-Million Naming Gift for Verizon Executive Education Center," Cornell Tech, February 2, 2015, https://tech.cornell.edu/news/cornell-tech-announces-50 -million-naming-gift-for-verizon-education-center (accessed May 11, 2017).

37. Picon, *Smart Cities*, 98.

38. Ibid., 13.

39. Cornell University, "Cornell Tech: Designed for Impact."

40. "Anti Development New Yorkers against the Cornell Technion Partnership," NYACT, last modified July 29, 2014, https://againstcornelltechnion.wordpress.com/ (accessed December 3, 2014).

41. Lange, "Silicon Island."

42. Cornell University, "Cornell Tech: Designed for Impact."

43. "Matrix," *English Oxford Living English Dictionaries*, https://en.oxforddictionaries.com/defini tion/matrix (accessed March 2, 2016).

44. Nancy Scola, "Tech and the City," *Next City*, September 3, 2012, https://nextcity.org/ forefront/view/Tech-and-the-city (accessed September 11, 2014).

45. "Mike Bloomberg's Record of Progress," *Mike*, n.d., https://mbprogress.connectionsmedia .com/jobs/technology-and-applied-sciences (accessed April 1, 2017).

46. Scola, "Tech and the City."

47. Ibid.

48. Lange, "Silicon Island."

49. Ibid.

50. "Michael Bloomberg," *Forbes*, n.d., http://www.forbes.com/profile/michael-bloomberg/; http://www.forbes.com/billionaires/list/#version:static (accessed May 2, 2017).

51. "The Giving Pledge," Wikipedia, n.d., https://en.wikipedia.org/wiki/The_Giving_Pledge (accessed May 4, 2017).

52. "About," Hudson Yards, http://www.hudsonyardsnewyork.com/ (accessed March 2, 2016).

53. Ibid.

54. Chris Smith, "Autocrat for the People," *New York Mag*, September 8, 2013, http://nymag .com/news/politics/bloomberg/legacy-2013-9/ (accessed January 30, 2016).

55. "NYU CUSP, Related Companies, and Oxford Properties Group Team Up to Create 'First Quantified Community' in the United States at Hudson Yards," NYU CUSP, April 14, 2014, http://cusp.nyu.edu/press-release/nyu-cusp-related-companies-oxford-properties-group-team -create-first-quantified-community-united-states-hudson-yards/ (accessed October 3, 2015).

56. Ibid.

57. Ibid.

58. Dominique Lorraine, "The Discrete Hand: Global Finance and the City," *Presses de Sciences Po* 61, trans. Sarah Louise Raillard (2011): 60, 15.

59. Agnès Sander, "Paquebots Urbains," *Flux* 50 (2002), http://olegk.free.fr/flux/Flux50/ Sommairefl50.html (accessed March 11, 2014).

60. Lorraine, "The Discrete Hand," 60.

61. Ibid., 63 and 69; See also Mike Davis and Daniel Bertrand Monk, eds., *Evil Empires: Dreamworlds of Neoliberalism* (New York: The New Press, 2008).

62. Martin Ford, *The Rise of the Robots: Technology and the Threat of Mass Unemployment* (London: Oneworld Publications, 2015), 35–36, Kindle edition.

63. Emmanuel Saez, "Striking it Richer: The Evolution of Top Incomes in the United States," *University of California, Berkeley*, September 3, 2013, https://eml.berkeley.edu/~saez/saez -UStopincomes-2012.pdf (accessed June 14, 2015).

64. Katia Savchuk, "New York Is the City with the Most Billionaires, Not Beijing," *Forbes*, March 1, 2016, https://www.forbes.com/sites/katiasavchuk/2016/03/01/new-york-most-billionaires/ (accessed March 3, 2016); a fact contested by Charles Reilly, "Beijing Now Has More Billionaires than New York," *CNN Money*, February 24, 2016, http://money.cnn.com/2016/02/24/investing/ beijing-new-york-billionaires/(accessed March 3, 2016); Also "Cities with Most Billionaires,"

World Atlas, February 9, 2017, https://www.worldatlas.com/articles/cities-boasting-the-most -millionaires-around-the-globe.html (accessed February 11, 2017).

65. Ford, *The Rise of the Robots*, 889–893, 1976–1983.

66. "Equinix Connects Hibernia Express Sub-Sea Cable between New York and London," *Equinix*, November 3, 2015, https://www.equinix.com/newsroom/press-releases/pr/123417/equinix-con nects-hibernia-express-sub-sea-cable-between-new-york-and-london/ (accessed May 1, 2017).

67. Jessica Leber, "New York's New $20 Billion Neighborhood of Skyscrapers Is Designed with Millennials in Mind," *Fast Company*, July 29, 2014, https://www.fastcoexist.com/3033355/new -yorks-new-neigh-20-billion-neigborhood-of-skyscrapers-is-designed-with-millennials-in-mind (accessed July 30, 2014).

68. Jamie Condliffe, "New York City Is Building for a Future of Flooding," *MIT Technology Review*, January 30, 2017, https://www.technologyreview.com/s/603527/new-york-city-is-building -for-a-future-of-flooding/ (accessed May 1, 2017); Also see Jamie Condliffe, "New York City Is Weighing Plans for Flood Defenses," *MIT Technology Review*, July 6, 2016, https://www.techno logyreview.com/s/601850/new-york-city-is-weighing-ambitious-plans-for-flood-defenses/ (accessed May 1, 2017).

69. Bratton, *The Stack*, 162.

70. Maurizio Lazzarato, "'Exiting Language,' Semiotic Systems and the Production of Subjectivity in Félix Guattari," in *Cognitive Architecture: From Bio-politics to Noo-politics*, trans. Eric Anglés, ed. Deborah Hauptmann and Warren Neidich (Rotterdam: 010 Publishers, 2010), 503.

71. Bratton, *The Stack*, 162.

72. Ibid., 148 and 149.

73. "34th Street–Hudson Yards," *Wikipedia*, https://en.wikipedia.org/wiki/34th_Street%E2% 80%93Hudson_Yards_(IRT_Flushing_Line (accessed May 2, 2017).

74. Annie Ju, "NYC Tech Campus Finds Temporary Home at Google Headquarters," *Ezra*, Cornell University, May 21, 2012, https://ezramagazine.cornell.edu/SUMMER12/NYC2.html (accessed May 1, 2017).

75. Wolfman-Arendt, "Creating an Ever-Flexible Center for Tech Innovation."

76. Nick Pinto, "Google Is Transforming NYC's Payphones into a 'Personalized Propaganda Engine,'" *Village Voice*, July 6, 2016, https://www.villagevoice.com/news/google-is-transforming -nycs-payphones-into-a-personalized-propaganda-engine-8822938 (accessed May 3, 2017).

77. Ibid.

78. Ibid.

79. "Open Data for All New Yorkers," *Open Data NYC*, 2017, https://opendata.cityofnewyork.us/ (accessed May 4, 2017).

80. Bratton uses this term in the context of Gobelki Tepe. I appropriate this terminology to draw parallels between New York and Google's predominant role and power (revenue) base. See Bratton, *The Stack*, 148.

81. "Coolhunting," *Wikipedia*, https://en.wikipedia.org/wiki/Coolhunting (accessed June 5, 2017).

82. Picon, *Smart Cities*, 44–46.

83. See Bratton, *The Stack*; See also "AbbVie and Calico Announce a Novel Collaboration to Accelerate the Discovery, Development, and Commercialization of New Therapies," *Calico News*, September 3, 2014, http://www.calicolabs.com/news/2014/09/03/ (accessed March 2, 2015); Anthony Regalado, "Is Google Cornering the Market on Deep Learning?," *MIT Technology Review*, January 29, 2014, https://www.technologyreview.com/news/524026/is-google-cornering-the-market-on-deep-learning/ (accessed February 1, 2014); Anthony Regalado, "Google Wants to Store Your Genome," *MIT Technology Review*, November 6, 2014, https://www.technologyreview.com/s/532266/google-wants-to-store-your-genome/ (accessed February 24, 2015); Anthony Regalado, "Internet of DNA," *MIT Technology Review*, February 8, 2015, https://www.technologyreview.com/s/535016/internet-of-dna/ (accessed February 24, 2015); Tess Ingram, "The Square Kilometer Array: Going to Infinity and Beyond," *The Australian Financial Review*, February 24, 2017, http://www.afr.com/technology/the-square-kilometer-array-going-to-infinity-and-beyond-20170222-guine8 (accessed May 4, 2017).

84. "Manufacturing Life with Craig. J Venter," University of California Television, November 1, 2012, https://www.youtube.com/watch?v=PKtozMvSsBk (accessed May 29, 2013).

85. Alon Halevy, Peter Norvig, and Fernando Pereira, "The Unreasonable Effectiveness of Data," *IEEE Intelligent Systems Journal* 24 (2009): 8–12. Incidentally, these authors all work for Google, including Peter Norvig, its head of Artificial Intelligence.

86. Viktor Mayer-Schönberger and Kenneth Cukier, *Big Data: A Revolution That Will Transform How We Live, Work, and Think* (New York: Houghton Mifflin Harcourt, 2013), 41.

87. Ibid., 57–58.

88. James Bailey, "Driven by Data, Your Bank Can Predict Your Divorce," *Forbes*, November 15, 2011, https://www.forbes.com/sites/techonomy/2011/11/15/driven-by-data-your-bank-can-predict-your-divorce/ (accessed July 12, 2013).

89. Mayer-Schönberger and Cukier, *Big Data*, 56.

90. Chris Anderson, "The End of Theory: The Data Deluge Makes the Scientific Method Obsolete," *Wired*, June 23, 2008, https://www.wired.com/2008/06/pb-theory/ (accessed March 11, 2013).

91. Ibid.

92. Ibid.

93. Jonathon Taplin, *"Move Fast and Break Things* Review–Google, Facebook and Amazon Exposed," *The Guardian*, April 17, 2017, https://www.theguardian.com/books/2017/apr/17/move -fast-and-break-things-review-google-facebook-amazon-exposed (accessed, April 17, 2017).

94. Pinto, "Google Is Transforming NYC."

95. Ian Sample, "Breakthrough Prize Awards $25m to Researchers at 'Oscars of Science,'" *The Guardian*, December 5, 2016, https://www.theguardian.com/science/2016/dec/05/breakthrough -prize-awards-2016-25m-to-researchers-at-oscars-of-science (accessed, December 12, 2017).

96. "After [Frederick] Terman made Stanford 'Stanford,' he was hired by a number of cities to write proposals on how to turn their regions into a Silicon Valley. Among others, Bell Labs and Princeton asked him how to build a new university in New Jersey that was going to emulate Silicon Valley. In his proposal, Terman recommended that they make it graduate only, engineering only, without any emphasis on departments, but with loose focus areas, and an emphasis on training in the local industries. That was in the '60s. Princeton and Bell Labs dropped the plan because of money issues. Today the mayor of NYC is doing exactly what Terman recommended." See Maria Teresa Cometto and Alessandro Piol, *Tech and the City: The Making of New York's Startup Community* (New York: Mirandola Press, 2013), 2246–2257, Kindle edition.

97. Félix Guattari, *Les Années d'Hiver* (Paris: Les Prairies Ordinaires, 2009), 128, quoted in Lazzarato, "'Exiting Language,'" 512.

98. Adrienne Lafrance, "The Internet Is Mostly Bots," *The Atlantic*, January 31, 2017, https://www.theatlantic.com/technology/archive/2017/01/bots-bots-bots/515043/ (accessed June 5, 2017).

99. Bratton, *The Stack*, 171 and 365.

100. Kelly considers the "technium," defined as "the greater global, massively interconnected system of technology vibrating around us" as somewhat messianically, and therefore teleologically, an emerging seventh kingdom of life. See Kevin Kelly, *What Technology Wants* (New York: Penguin, 2010), 11, 6. "Inevitably" refers to Kelly's latest book, *The Inevitable: Understanding the 12 Technological Forces That Will Shape Our Future* (New York: Viking, 2016).

101. *Google and the World Brain*, directed by Ben Lewis (Barcelona: Polar Star Films, 2013), digital.

102. James Temple, "More on DeepMind: AI Startup to Work Directly with Google's Search Team," *Recode*, January 27, 2014, https://www.recode.net/2014/1/27/11622778/more-on-deep mind-ai-startup-to-work-directly-with-googles-search-team (accessed February 12, 2014).

Selected Bibliography

Abella, Alex. *Soldiers of Reason: The RAND Corporation and the Rise of American Empire*. New York: Harcourt, 2008.

Adkins, Lisa, and Celia Lury. "The Labour of Identity: Performing Identities, Performing Economies." *Economy and Society* 28 (1999): 598–614.

Aldrin, Edwin (Buzz) Jr., with Ken Abraham. *Magnificent Desolation: The Long Journey Home from the Moon*. New York: Crown Archetype, 2009.

Aleksandrov, V., and Zolotov, D. "'Tak derzhat!'" *Krokodil* 22, August 1983.

Allen, John. *Biosphere 2: The Human Experiment*. New York: Penguin Books, 1991.

Allen, John, and Mark Nelson. *Space Biospheres*. Malabar, FL: Orbit Book Company, Inc, 1987.

Allen, John, Tango Parrish, and Mark Nelson. "The Institute of Ecotechnics: An Institute Devoted to Developing the Discipline of Relating Technosphere to Biosphere." *The Environmentalist* 4 (1984): 205–218.

Allen, Richard Sanders. *The Northrop Story, 1929–1939*. New York: Orion, 1990.

Allen, Thomas J. *Managing the Flow of Technology*. Cambridge, MA: MIT Press, 1977.

Altman, Yochanan, and Yehuda Baruch. "The Organizational Lunch." *Culture and Organization* 16, no. 2 (2010): 127–143.

Anderson, Chris. "The End of Theory: The Data Deluge Makes the Scientific Method Obsolete." *Wired*, June 23, 2008. https://www.wired.com/2008/06/pb-theory/.

Anker, Peder. "The Closed World of Ecological Architecture." *Environmental History* 10, no. 2 (April 2005): 527–552.

Anker, Peder. "The Ecological Colonization of Space." *Environmental History* 10, no. 2 (April 2005): 239–268.

Appenzeller, Tim. "Biosphere 2 Makes a New Bid for Scientific Credibility." *Science* 263, no. 5152 (March 11, 1994): 1368–1369.

Bakhtin, Mikhail. *Rabelais and His World*. Bloomington: Indiana University Press, [1965] 1984.

Barthes, Roland. *Writing Degree Zero*. London: Jonathan Cape, [1953] 1967.

Baumgardt, Carola. *Johannes Kepler Life and Letters*. Introduction by Albert Einstein. New York: Philosophical Library, 1951.

Beardsley, Tim. "Down to Earth: Biosphere 2 Tries to Get Real." *Scientific American* (August 1995).

Bednar, Michael J. *The New Atrium*. New York: McGraw-Hill, 1986.

Beers, David. *Blue Sky Dream: A Memoir of America's Fall from Grace*. New York: Doubleday, 1996.

Bentham, Jeremy. *The Panopticon Writings*, ed. Miran Bozovic. London: Verso, 1995.

Bentham, Jeremy. *The Works of Jeremy Bentham*, vol. 4 (Panopticon, Constitution, Colonies, Codification), ed. John Bowring (1843). Online Library of Liberty, Liberty Fund. One of eleven volumes, orig. published by William Tait, Edinburgh. http://oll.libertyfund.org/titles/1925.

Berg, Raissa. *Acquired Traits: Memoirs of a Geneticist from the Soviet Union*. New York: Viking Penguin, 1988.

Berkowitz, Eric, et al. *Marketing*. Homewood, IL: Irwin, 1992.

Bimm, Jordan. "Rethinking the Overview Effect." *Quest: The History of Spaceflight Quarterly* 21, no. 1 (2014): 39–47.

Boltanski, Luc, and Eve Chiapello. *The New Spirit of Capitalism*. Trans. G. Elliot. London: Verso, 2017.

Bolton, Jonathan. *Worlds of Dissent: Charter 77, The Plastic People of the Universe, and Czech Culture under Communism*. Cambridge, MA: Harvard University Press, 2012.

Borzenkov, A. G. *Molodezh´ i Politika: Vozmozhnosti i Predely Studencheskoi Samodeiatel'nosti na vostoke Rossii (1961–1991 gg.)*, 3 vols. Novosibirsk: Novosibirskii Gosudarstvennyi Universitet, 2003.

Bourdieu, Pierre. *Distinction: A Social Critique of the Judgment of Taste*. Trans. Richard Nice. Cambridge, MA: Harvard University Press, [1979] 1984.

Brand, Stewart. "SPACEWAR: Fanatic Life and Symbolic Death among the Computer Bums." *Rolling Stone*, December 7, 1972.

Braseth, Timothy. "Josef van der Kar: Building Architectural Bridges." *Modernism* 14, no. 2 (Summer 2011): 44–53.

Bratton, Benjamin. *The Stack: On Software and Sovereignty*. Cambridge, MA: MIT Press, 2015.

Brown, Kate. *Plutopia: Nuclear Families, Atomic Cities, and the Great Soviet and American Plutonium Disasters*. Oxford: Oxford University Press, 2013.

Brullet, Manuel, Albert de Pineda, and Alfonso de Luna. *Parc de Recerca Biomèdica de Barcelona (PRBB)*. Barcelona: Edicions de l'Eixample, 2007.

Burshtein, A. I. "Ot 'goroda Solntsa' k gorodu 'Zero.'" In *I zabyt' po-prezhnemu nel'zia*, ed. N. A. Pritvits and S. P. Rozhnova, 127–135. Novosibirsk: 2007.

Callon, Michel. "Concevoir: Modèle Hiérarchique et Modèle Négocié." In *L'Élaboration des Projets Architecturaux et Urbains en Europe*, ed. Michel Bonnet, 169–174. Paris: Plan Construction et Architecture, 1997.

Callon, Michel. "Le Travail de Conception en Architecture." *Situations Les Cahiers de la Recherche Architecturale* 37 (1996): 25–35.

Campbell, Virginia. "How RAND Invented the Postwar World." *American Heritage of Invention & Technology* 20 (Summer 2004): 50–59.

Canguilhem, Georges. *Ideology and Rationality in the History of the Life Sciences.* Trans. A. Goldhammer. Cambridge, MA: MIT Press, 1988.

Cederström, Carl, and André Spicer. *The Wellness Syndrome.* Cambridge, UK: Polity Press, 2015.

Clareson, Thomas D. *Science Fiction Criticism: An Annotated Checklist.* The Serif Series: Bibliographies and Checklists. Kent, OH: Kent State University Press, 1972.

Cohn, Jeffrey P. "Biosphere 2, Version 3.0." *Bioscience* 57, no. 9 (October 2007): 808.

Collins, Martin. *Cold War Laboratory: RAND, the Air Force, and the American State, 1945–1950.* Washington, DC: Smithsonian Institution Scholarly Press, 2002.

Cometto, Maria Teresa, and Alessandro Piol. *Tech and the City: The Making of New York's Startup Community.* New York: Mirandola Press, 2013. Kindle edition.

Corsín Jiménez, Alberto, and Rane Willerslev. "'An Anthropological Concept of the Concept': Reversibility among the Siberian Yukaghirs." *Journal of the Royal Anthropological Institute* 13 (2007): 527–544.

Cressy, David. "Early Modern Space Travel and the English Man in the Moon." *American Historical Review* 111, no. 4 (October 2006): 961–982.

Crosbie, Michael. "Add and Subtract." *Progressive Architecture* 74, no. 10 (October 1993): 48–51.

Cuff, Dana. *Architecture: The Story of Practice.* Cambridge, MA: MIT Press, 1991.

Davenport, Thomas. *Thinking for a Living: How to Get Better Performances and Results from Knowledge Workers.* Boston: Harvard Business School Press, 2005.

Davis, Mike, and Daniel Bertrand Monk, eds. *Evil Empires: Dreamworlds of Neoliberalism.* New York: The New Press, 2008.

de Bono, Edward. *The Use of Lateral Thinking.* New York: Basic Books, 1968.

Deleuze, Gilles. "Postscript on the Societies of Control." *October* 59 (1992): 3–7.

Deleuze, Gilles, and Félix Guattari. *Anti-Oedipus: Capitalism and Schizophrenia.* Trans. R. Hurley, M. Seem, and H. R. Lane. Minneapolis: University of Minnesota Press, 1983.

Dempster, William, and Mark Nelson. "Living in Space: Results from Biosphere 2's Initial Closure, an Early Testbed for Closed Ecological Systems of Mars." In *Strategies for Mars: A Guide to Human Exploration, Science and Technology Series*, ed. Carol R. Stoker and Carter Emmart, 373–390. San Diego, CA: American Astronautical Society, 1996.

Derrida, Jacques. *The Animal That Therefore I Am*. New York: Fordham University Press, 2008.

Derrida, Jacques. *Dissemination*. Trans. B. Johnson. Chicago: Chicago University Press, [1972] 1981.

Derrida, Jacques. *Of Grammatology*. Trans. G. C. Spivak. Baltimore: Johns Hopkins University Press, [1967] 1976.

Derrida, Jacques. *Positions*. Trans. A. Bass. Chicago: Chicago University Press, [1972] 1981.

Derrida, Jacques. *Writing and Difference*. Trans. A. Bass. London: Routledge & Kegan Paul, [1967] 1978.

D'Hooghe, Alexander. "Science Towns as Fragments of a New Civilization: The Soviet Development of Siberia." *Interdisciplinary Science Reviews* 31 (2) (2006): 135–148.

Dierig, Sven, Jens Lachmund, and Andrew Mendelsohn, eds. "Science and the City." Special issue, *Osiris* 18 (2003).

Dobbs, Richard, James Manyika, and Jonathan Woetzel. *No Ordinary Disruption: The Four Global Forces Breaking All the Trends*. New York: Perseus, 2015. Kindle edition.

Dobers, Peter, and Lars Strannegård. "Design, Lifestyles and Sustainability: Aesthetic Consumption in a World of Abundance." *Business Strategy and the Environment* 14 (2005): 325–326.

Duffy, Francis, and Jack Tanis. "A Vision of the New Workplace." *Site Selection and Industrial Development* 162 (April 1993): 428.

du Gay, Paul. *Consumption and Identity at Work*. London: Sage, 1996.

Eikhof, Doris, Chris Warhurst, and Axel Haunschild. "Introduction: What Work? What Life? What Balance? Critical Reflections on the Work-Life Balance Debate." *Employee Relations* 29 (2007): 325–333.

Ensmenger, Nathan. "Beards, Sandals, and Other Signs of Rugged Individualism: Masculine Culture within the Computing Professions." *Osiris* 30, no. 1 (2015): 38–65.

Eugenides, Jeffrey. *The Marriage Plot*. New York: Farrar, Straus and Giroux, 2011.

Evans, Arthur B. "The Origins of Science Fiction Criticism: From Kepler to Wells." *Science Fiction Studies* 26, no. 2 (1999): 163–186.

Evans, Christine Elaine. "From Truth to Time: Soviet Central Television, 1957–1985." PhD diss., UC Berkeley, 2010.

Featherstone, Mike. *Consumer Culture and Postmodernism*. London: Sage, 1991.

Featherstone, Mike. "Leisure, Symbolic Power and the Life Course." In *Sport, Leisure and Social Relations* (Sociological Review Monograph 33), ed. John Horne, David Jary, and Alan Tomlinson, 113–138. London: Routledge and Kegan Paul, 1987.

Featherstone, Michael. "Luxury, Consumer Culture and Sumptuary Dynamics." *Luxury* 1 (2014): 47–69.

Florida, Richard. *The Rise of the Creative Class*. New York: Basic Books, 2002.

Florida, Richard. *The Rise of the Creative Class, Revisited*. New York: Basic Books, 2012.

Ford, Martin. *The Rise of the Robots: Technology and the Threat of Mass Unemployment*. London: Oneworld Publications, 2015. Kindle edition.

Foucault, Michel. *Archaeology of Knowledge*. Trans. A. M. Sheridan-Smith. New York: Pantheon Books, [1969] 1972.

Foucault, Michel. *The History of Sexuality: The Will to Knowledge*. Vol. 1. Trans. Robert Hurley. London: Penguin, 2008.

Foucault, Michel. "Preface." In Gilles Deleuze and Félix Guattari, *Anti-Oedipus: Capitalism and Schizophrenia*, trans. R. Hurley, M. Seem, and H. R. Lane, xi–xiv. Minneapolis: University of Minnesota Press, 1983.

Frampton, Kenneth, and Steven Moore. "Technology and Place." *Journal of Architectural Education* 54, no. 3 (2001): 121–122.

Frederick, Christine. *The New Housekeeping: Efficiency Studies in Home Management*. Garden City, NY: Doubleday, Page & Company, 1918.

Gagliardi, Pasquale. *Symbols and Artifacts: Views of the Corporate Landscape*. Berlin: Walter de Greyter, 1990.

Galison, Peter. *Image and Logic: A Material Culture of Microphysics*. Chicago: University of Chicago Press, 1997.

Galison, P., and E. Thompson, eds. *The Architecture of Science*. Cambridge, MA: MIT Press, 1999.

Gaponov, Iu. V. "Traditsii 'fizicheskogo iskusstva' v rosiiskom fizicheskom soobchshestve 50-90kh godov." *Voprosy istorii estestvoznaniia i tekhniki* 4 (2003): 165–178.

Garey, Amy. "Aleksandr Galich: Performance and the Politics of the Everyday." *Limina: A Journal of Historical and Cultural Studies* 17 (2011): 1–13.

Garipov, R. M. "Po stsenariiu Lavrentieva." In *Vek Lavrentieva*, ed. N. A. Pritvits, V. D. Ermikov, and Z. M. Ibragimova, 181–285. Novosibirsk: Zdatel'stvo SO RAN, 2000.

Genter, Robert. "With Great Power Comes Great Responsibility: Cold War Culture and the Birth of Marvel Comics." *Journal of Popular Culture* 40 (2007): 953–978.

Gertner, Jon. *The Idea Factory: Bell Labs and the Great Age of American Innovation*. London: Penguin Books, 2012.

Ghamari-Tabrizi, Sharon. *The Worlds of Herman Kahn: The Intuitive Science of Thermonuclear War*. Cambridge, MA: Harvard University Press, 2005.

Gieryn, Thomas. "City as Truth-Spot: Laboratories and Field-Sites in Urban Studies." *Social Studies of Science* 36, no. 1 (2006): 5–38.

Gieryn, Thomas. "Two Faces on Science: Building Identities for Molecular Biology and Biotechnology." In *The Architecture of Science*, ed. Peter Galison and Emily Thompson, 423–459. Cambridge, MA: MIT Press, 1999.

Gieryn, Thomas. "What Buildings Do." *Theory and Society* 31, no. 1 (February 2002): 35–74.

Goodman, Donna. *A History of the Future*. New York: Monacelli Press, 2008.

Gordin, Michael, Helen Tilley, and Gyan Prakash. *Utopia/Dystopia: Conditions of Historical Possibility*. Princeton, NJ: Princeton University Press, 2010.

Gorsuch, Anne E., and Diane P. Koenker. *Turizm: The Russian and East European Tourist under Capitalism and Socialism*. Ithaca, NY: Cornell University Press, 2006.

Grattan-Guiness, Ivor. "The 'Ingeneur Savant,' 1800–1830: A Neglected Figure in the History of French Mathematics and Science." *Science in Context* 6, no. 2 (1993): 405–433.

Gray, Kathelin. "A Long-Duration Art Ensemble: Theatre of All Possibilities, Research Vessel Heraclitus, and Biosphere 2: A Dancer's Perspective." In *The Ethics of Art Ecological Turns in the Performing Arts*, ed. Guy Cools and Pascal Gielen, 159–185. Amsterdam: Valiz, 2014.

Grinevald, Jacques. "Sketch for a History of the Idea of the Biosphere." In *Gaia in Action, Science of the Living Earth*, ed. P. Bunyard, 34–53. Edinburgh: Floris Books, 1996.

Grove, Richard. *Green Imperialism: Colonial Expansion, Tropical Island Edens, and the Origins of Environmentalism, 1600–1860*. Cambridge, Melbourne: Cambridge University Press, 1996.

Gruen, Victor, and Larry Smith. *Shopping Towns USA: The Planning of Shopping Centers*. New York: Reinhold Publishing, 1960.

Guilbaut, Serge. *How New York Stole the Idea of Modern Art*. Chicago: University of Chicago Press, 1985.

Gunn, W., T. Otto, and R. C. Smith, eds. *Design Anthropology: Theory and Practice*. London: Bloomsbury Academic, 2013.

Halevy, Alon, Peter Norvig, and Fernando Pereira. "The Unreasonable Effectiveness of Data." *IEEE Intelligent Systems Journal* 24 (2009): 8–12.

Hall, Peter. *Cities of Tomorrow*. Oxford: Blackwell, 2002.

Halpin, Helen Ann, Maria M. Morales-Suárez-Varela, and José M. Martin-Moreno. "Chronic Disease Prevention and the New Public Health." *Public Health Reviews* 32 (2010): 120–154.

Hamilton, Andrew. "Shangri-La for Deep Thinkers." *Westways* 54 (February 1962): 4–6.

Hancock, Philip. "Uncovering the Semiotic in Organizational Aesthetics." *Organization* 12 (2005): 30.

Haraway, Donna J. *When Species Meet*. Minneapolis: University of Minnesota Press, 2008.

Harris, William C., and Lisa J. Graumlich. "Biosphere 2: Sustainable Research for a Sustainable Planet." *21st Century* 4, no. 1 (n.d.) http://www.columbia.edu/cu/21stC/issue-4.1/harris.html.

Harvey, David. *Spaces of Hope*. Berkeley: University of California Press, 2000.

Heidegger, Martin. "Building Dwelling Thinking." In *Poetry, Language, Thought*, 141–160. Trans. and introduction Albert Hofstadter. New York: Harper & Row, 1975.

Helmreich, Stephan. "What Was Life? Answers from Three Limit Biologies." *Critical Inquiry* 37 (2011): 671–696.

Hillier, Bill, and Allen Penn. "Visible Colleges: Structure and Randomness in the Place of Discovery." *Science in Context* 4 (1991): 23–49.

Hiltzik, Michael A. *Dealers of Lightning: Xerox PARC and the Dawn of the Computer Age*. New York: Harper, 2000.

Hirschman, Elizabeth, Linda Scott, and William Wells. "A Model of Product Discourse: Linking Consumer Practice to Cultural Texts." *Journal of Advertising* 27 (1998): 33–50.

Höhler, Sabine. "The Environment as a Life Support System: The Case of Biosphere 2." *History and Technology* 26, no. 1 (March 2010): 39–58.

Houdart, Sophie, and Minato Chihiro. *Kuma Kengo: An Unconventional Monograph*. Paris: Editions Donner Lieu, 2009.

Hughes, Russell. "The Internet of Politicized 'Things': Urbanization, Citizenship, and the Hacking of New York 'Innovation' City." *Interstices Journal of Architecture and Related Arts* 16 (2016): 24–30.

Hunner, Jon. *Inventing Los Alamos: The Growth of an Atomic Community*. Norman: University of Oklahoma Press, 2007.

Huxley, Margo. "Geographies of Governmentality." In *Space Knowledge and Power: Foucault and Geography*, ed. Jeremy Crampton and Stuart Elden, 185–204. Hampshire: Ashgate, 2007.

Hyland, L. A. *Call Me Pat: The Autobiography of the Man Howard Hughes Chose to Lead Hughes Aircraft*. Virginia Beach, VA: Donning Company, 1993.

Ibragimova, Z. M., and N. A. Pritvits. *Treugol'nik Lavrentieva*. Moskva: Sovetskaia Rossiia, 1989.

Ilatovskaia, T. "V poiskakh 'sumasshedshei' idei." *Smena* 840, May 1962.

Il'f, Iliia. *Iz zapisnykh knizhek*. Leningrad: Khudozhnik RSFSR, 1966.

Ingraham, Catherine. *Architecture, Animal, Human: The Asymmetrical Condition*. London: Routledge, 2006.

James, Nancy. "Realism in Romance: A Critical Study of The Short Stories of Edward Everett Hale." PhD diss., Pennsylvania State University, 1969.

Johnston, Josée, and Shyon Baumann. *Foodie: Democracy and Distinction in the Gourmet Foodscape.* New York: Routledge, 2015.

Josephson, Paul. *New Atlantis Revisited: Akademgorodok, The Siberian City of Science.* Princeton, NJ: Princeton University Press, 1997.

Josephson, Paul. *Physics and Politics in Revolutionary Russia.* Berkeley: University of California Press, 1991.

Kaiser, David. "The Postwar Suburbanization of American Physics." *American Quarterly* 56, no. 4 (2004): 851–888.

Kaji-O'Grady, Sandra. "Laboratories of Experimental Science." In *The Architecture of Industry: Changing Paradigms in Industrial Building and Planning,* ed. Mathew Aitchison, 109–135. Hampshire: Ashgate, 2014.

Kaji-O'Grady, Sandra. "The Spaces of Experimental Science." *Le Journal Spéciale'Z* 4 (2012): 88–99.

Kaji-O'Grady, Sandra, and Chris L. Smith. "Exaptive Translations between Biology and Architecture." *Architecture Research Quarterly* 18, no. 2 (2014): 155–166.

Kaji-O'Grady, Sandra, and Chris L. Smith. "Laboratory Architecture and the Deep Skin of Science." In *Industries of Architecture,* ed. Katie Lloyd Thomas, Nick Beech, and Tilo Amhoff, 282–293. London: Routledge, 2015.

Kaplan, Fred. *The Wizards of Armageddon.* New York: Simon and Schuster, 1983.

Karlov, N. V. *Povest' drevnikh vremen, ili predistoriia Fiztekha.* Moscow: MTsGO MFTI, 2005.

Katz, Barry M. "Research and Development." In *Make It New: The History of Silicon Valley Design.* Cambridge, MA: MIT Press, 2015. http://common.books24x7.com.libezproxy2.syr.edu/toc.aspx?bookid=112067.

Kaufmann-Buhler, Jennifer. "From the Open Plan to the Cubicle: The Real and Imagined Transformation of American Office Design and Office Work, 1945–1999." PhD diss., University of Wisconsin-Madison, 2013.

Kelly, Kevin. *The Inevitable: Understanding the 12 Technological Forces That Will Shape Our Future.* New York: Viking, 2016.

Kelly, Kevin. *Out of Control: The New Biology of Machines, Social Systems and the Economic World.* Boston: Addison-Wesley, 1994.

Kelly, Kevin. *What Technology Wants.* New York: Penguin, 2010.

Kepler, Johannes. *Somnium (The Dream).* 1608. The Somnium Project, 2011–2017. Trans. Tom Metcalf. https://somniumproject.wordpress.com/somnium/.

Kepler, Johannes, and Edward Rosen. *Kepler's Conversation with Galileo's Sidereal Messenger*. Trans. and ed Edward Rosen, The Sources of Science, no. 5. New York and London: Johnson Reprint Corp., 1965. http://digitalcollections.library.cmu.edu/awweb/awarchive?type=file&item=393654.

Kim, Janette, and Erik Carver. "Crisis in Crisis: Biosphere 2's Contested Ecologies." *Volume 20: Storytelling*, 2009. http://c-lab.columbia.edu/0167.html.

Knorr-Cetina, Karin D. *The Manufacture of Knowledge: An Essay on the Constructivist and Contextual Nature of Science*. Oxford: Pergamon, 1981.

Kubo, Michael. "Constructing the Cold War Environment: The Strategic Architecture of RAND." M.Arch thesis, Harvard University Graduate School of Design, 2009.

Kuchner, Marc. *Marketing for Scientists: How to Shine in Tough Times*. Washington, DC: Island Books, 2012.

Kuznetsov, I. S. *Pis'mo soroka shesti. Dokumental'noe izdanie*. Novosibirsk: Klio, 2007.

Landau, L. D., and E. M. Lifshits. *Mekhanika*. Moscow: Fizmatgiz, 1958.

Lawlor, Leonard. *This Is Not Sufficient: An Essay on Animality and Human Nature in Derrida*. New York: Columbia University Press, 2007.

Latour, Bruno. *Aramis, or the Love of Technology*. Trans. C. Porter. Cambridge, MA: Harvard University Press, 1996.

Latour, Bruno. *Science in Action: How to Follow Engineers through Society*. Milton Keynes: Open University Press, 1987.

Latour, Bruno, and Peter Weibel. *Making Things Public: Atmospheres of Democracy*. Cambridge, MA: MIT Press, 2005.

Latour, Bruno, and Steve Woolgar. *Laboratory Life: The Construction of Scientific Facts*. Princeton, NJ: Princeton University Press, 1986.

Latour, Bruno, and Steve Woolgar. *Laboratory Life: The Social Construction of Scientific Facts*. Beverly Hills, CA: Sage Publications, 1979.

Latour, Bruno, and Albena Yaneva. "Give Me a Gun and I Will Make All Buildings Move: An ANT's View of Architecture." In *Explorations in Architecture: Teaching, Design, Research*, ed. Reto Geiser, 80–89. Basel: Birkhäuser, 2008.

Lavrentiev, M. A. "Opyty zhizni." In *Vek Lavrentiev*, ed. N. A. Pritvits, V. D. Ermikov, and Z. M. Ibragimova, 15–375. Novosibirsk: Zdatel'stvo SO RAN, 2000.

Lavrentiev, Mikhail A. "Razvitie nauki v Sibiri i na Dal'nem Vostoke." *Vestnik AN SSSR* 12 (1957): 3–7.

Lazzarato, Maurizio. "'Exiting Language,' Semiotic Systems and the Production of Subjectivity in Félix Guattari." Trans. Eric Anglès. In *Cognitive Architecture: From Biopolitics to Noopolitics: Architecture and Mind in the Age of Communication and Information*, ed. Deborah Hauptmann and Warren Neidich, 502–520. Rotterdam: 010 Publishers, 2010.

Lazzarato, Maurizio. "Immaterial Labor." Trans. P. Colilli and E. Emery. In *Radical Thought in Italy: A Potential Politics*, ed. Paolo Virno and Michael Hardt, 133–147. Minneapolis: University of Minnesota Press, 1996.

Light, Jennifer. *From Warfare to Welfare: Defense Intellectuals and Urban Problems in Cold War America*. Cambridge, MA: MIT Press, 2003.

Livingston, David. *Putting Science in Its Place: Geographies of Scientific Knowledge*. Chicago: University of Chicago Press, 2003.

Logsdon, J. M., L. J. Lear, J. Warren-Findley, R. Williamson, and D. A. Day, eds. *Exploring the Unknown: Selected Documents in the History of the U.S. Civilian Space Program*, vol. 1. Washington, DC: National Aeronautics and Space Administration, NASA History Division, Office of Policy and Plans, 1998.

Lorraine, Dominique. "The Discrete Hand: Global Finance and the City." Trans. Sarah Louise Raillard. *Presses de Sciences Po* 61 (2011): 60. doi:10.3917/rfsp.616.1097.

Lotchin, Roger W. *Fortress California, 1910–1961: From Warfare to Welfare*. New York: Oxford University Press, 1992.

Luke, Timothy W. "Environmental Emulations: Terraforming Technologies and the Tourist Trade at Biosphere 2." In *Ecocritique: Contesting the Politics of Nature, Economy, and Culture*, 95–114. Minneapolis: University of Minnesota Press, 1997.

Lynch, Michael. *Art and Artifact in Laboratory Science: A Study of Shop Work and Shop Talk in a Research Laboratory*. Boston: Routledge and Kegan Paul, 1985.

Lynch, Michael. *Scientific Practice and Ordinary Action: Ethnomethodology and Social Studies of Science*. Cambridge: Cambridge University Press, 1993.

Lyons, Dan. *Disrupted: My Misadventure in the Start-Up Bubble*. New York: Hachette Books, 2016.

Lyotard, Jean-François. *The Postmodern Condition: A Report on Knowledge*. Trans. G. Bennington and B. Massumi. Minneapolis: University of Minnesota Press, 1993.

Machlup, Fritz. *The Production and Distribution of Knowledge in the United States*. Princeton, NJ: Princeton University Press, 1962.

Maiman, Theodore. *The Laser Odyssey*. Fairfield, CA: Laser Press, 2001.

Mallet, Marie-Louise. "Foreword." In *The Animal That Therefore I Am*, ed. Jacques Derrida, x–xi. New York: Fordham University Press, 2008.

Marine, Gene. "Think Factory De Luxe." *The Nation*, February 14, 1959, 131–135.

Markoff, John. *What the Dormouse Said: How the Sixties Counter-Culture Shaped the Personal Computer Industry*. London: Penguin Books, 2005.

Martin, Reinhold. "Architecture's Image Problem: Have We Ever Been Postmodern?," *Grey Room* 22 (Winter 2006): 6–29.

Mau, Bruce. *Life Style*. Ed. Kyo Maclear. New York: Phaidon, 2000.

Mayer-Schönberger, Viktor, and Kenneth Cukier. *Big Data: A Revolution That Will Transform How We Live, Work, and Think*. New York: Houghton Mifflin Harcourt, 2013.

McCoy, Esther. "Dr. Salk Talks about His Institute." *Architectural Forum* 127 (December 1967): 32.

McCray, W. Patrick. *The Visioneers: How a Group of Elite Scientists Pursued Space Colonies, Nanotechnologies, and a Limitless Future*. Princeton, NJ: Princeton University Press, 2012.

McElheny, Victor K. *Watson and DNA: Making a Scientific Revolution*. New York: Perseus/Wiley, 2003.

McKinlay, Alan, and Philip Taylor. Power, Surveillance and Resistance: Inside the 'Factory of the Future.'" In *The New Workplace and Trade Unionism*, ed. P. Ackers, C. Smith, and P. Smith, 279–300. London: Routledge, 1996.

Meshkov, I. N. "Tri istochnika i tri sostavnye chasti." In *Vek Lavrentieva*, ed. N. A. Pritvits, V. D. Ermikov, and Z. M. Ibragimova, 324–326. Novosibirsk: Zdatel'stvo SO RAN, 2000.

Mindolin, V. A. "Nostal'giia." In *I zabyt' po-prezhnemu nel'zia*, ed. N. A. Pritvits and S. P. Rozhnova, 169–174. Novosibirsk: IEOPP SO RAN, 2007.

Moehring, Eugene. "The Sahara Hotel: Las Vegas' Jewel in the Desert." In *Stripping Las Vegas: A Contextual Review of Casino Resort Architecture*, ed. Karin Jaschke and Silke Otsch, 13–39. London: Verso, 2004.

Morton, Timothy. 'This Is Not My Beautiful Biosphere.' In *A Cultural History of Climate Change*, ed. Tom Bristow and Thomas H. Ford, 229–238. London, UK: Routledge, 2016.

Moskowitz, Daniel B. "The Bucks behind the Wellness Boom." *Business and Health* 17, no. 2 (1999): 43–44.

Nathans, Benjamin. "Talking Fish: On Soviet Dissident Memoirs." *Journal of Modern History* 87 (2015): 579–614.

Neff, Gina. *Venture Labor: Work and the Burden of Risk in Innovative Industries*. Cambridge, MA: MIT Press, 2012.

Nelson, Mark. *Pushing Our Limits: Insights from Biosphere 2*. Tucson: Arizona University Press, 2017.

Newkirk, Roland W, Ivan D. Ertel, and Courtney G. Brooks. *Skylab: A Chronology*. Washington, DC: Scientific and Technical Information Office, National Aeronautics and Space Administration, 1977. https://history.nasa.gov/SP-4011/contents.htm.

Noordung, Hermann (Hermann Potočnik). *The Problem of Space Travel: The Rocket Motor*. Ed. Ernst Stuhlinger and J. D. Hunley with Jennifer Garland. Washington, DC: National Aeronautics and Space Administration, [1929] 1995. https://history.nasa.gov/SP-4026.pdf.

Odum, Eugene P. "A New Kind of Science." *Science* 260, no. 5110 (May 14, 1993): 878–879.

Odum, Howard T. "Scales of Ecological Engineering." *Ecological Engineering* 6 (1996): 7–19.

Otto, Ton, and Rachel Charlotte Smith. "Design Anthropology: A Distinct Style of Knowing." In *Design Anthropology: Theory and Practice*, ed. Wendy Gunn, Ton Otto, and Rachel Charlotte Smith, 1–29. London: Bloomsbury Academic, 2013.

Palmer, S., ed. *Architecture at Work: DMJM Design Los Angeles*. New York: Edizioni Press, 2004.

Papers of John F. Kennedy. Presidential Papers. National Security Files. Subjects. Space activities: Long Range Plans of NASA (National Aeronautics and Space Administration), vol. I–IV, May 29, 1962, 10–11. https://www.jfklibrary.org/Asset-Viewer/Archives/JFKNSF-307-002.aspx.

Papers of John F. Kennedy. Presidential Papers. President's Office Files. Speech Files. Address at Rice University, Houston, Texas, September 12, 1962, 3. https://www.jfklibrary.org/Asset-Viewer/Archives/JFKPOF-040-001.aspx.

Papert, Seymour. "Teaching Children Thinking." In *Proceedings of IFIPS World Congress on Computers and Education*. Amsterdam: North-Holland, 1972.

Peters, Benjamin. *How Not to Network a Nation, The Uneasy History of the Soviet Internet*. Cambridge, MA: MIT Press, 2016.

Pickering, Andrew. *The Cybernetic Brain: Sketches of Another Future*. Chicago, IL: University of Chicago Press, 2009.

Pickering, Andrew. *Science as Practice and Culture*. Chicago: University of Chicago Press, 1992.

Picon, Antoine. *Smart Cities: A Spatialised Intelligence*. Chichester: Wiley, 2015.

Picon, A., and A. Ponte, eds. *Architecture and the Sciences: Exchanging Metaphors*. New York: Princeton Architectural Press, 2003.

Pospelov, G. V. "Bol'shaia nauka v Sibiri." *Sibirskie Ogni* 12 (1959): 159–170.

Pospelov, G. V. "V nauchnom tsentre Sibiri." *Nash sovremennik* 3 (1958): 319–321.

Prashkevich, G. M., and T. A. Ianushevich. *Zelenoe vino: Literaturnyi Akademgorodok shestidesiatykh*. Novosibirsk: Svin'in i Synov'ia, 2009.

Prashkevich, G. M., T. A. Ianushevich, and V. Svin'in. *Shkola geniev*. Novosibirsk: Nashe nasledie Moskova, 2004.

Pulos, Arthur. *The American Design Adventure, 1940–1975*. Cambridge, MA: MIT Press, 1988.

Raizman, David. *History of Modern Design: Graphics and Products since the Industrial Revolution*. London: Laurence King Publishing, 2003.

Ramo, Simon. *The Business of Science: Winning and Losing in the High-Tech Age*. New York: Hill and Wang, 1988.

Reider, Rebecca. *Dreaming the Biosphere: The Theater of All Possibilities*. Albuquerque: University of New Mexico Press, 2009.

Richard, Analiese, and Daromir Rudnyckyj. "Economies of Affect." *Journal of the Royal Anthropological Institute* 15, no. 1 (2009): 57–77.

Richards, Greg. "Evolving Gastronomic Experiences: From Food to Foodies to Foodscapes." *Journal of Gastronomy and Tourism* 1 (2015): 5–17.

Richardson, D. Kenneth. *Hughes after Howard: The Story of Hughes Aircraft Company*. Santa Barbara, CA: Sea Hill Press, 2011.

Roach, Mary. *Packing for Mars: The Curious Science of Life in the Void*. New York: W. W. Norton & Company, 2010.

Ross, Andrew. *No-Collar: The Humane Workplace and Its Hidden Costs*. New York: Basic Books, 2003.

Russell, James S. "Form Follows Fad: The Troubled Love Affair of Architectural Style and Management Ideal." In *On the Job: Design and the American Office*, ed. Donald Albrecht and Chrysanthe B. Broikos, 49–73. Princeton, NJ: Princeton Architectural Press, 2000.

Sadler, Simon. "An Architecture of the Whole." *Journal of Architectural Education* 61, no. 4 (May 2008): 108–129.

Sagan, Carl, and Adrian Malone. *Cosmos*. Collector's ed. Studio City, CA: Cosmos Studios, episode 3, 2000.

Sander, Agnès. "Paquebots urbains." *Flux* 50 (2002). http://olegk.free.fr/flux/Flux50/Sommairefl50.html.

Schmidgen, Henning, and Gloria Custance. *Bruno Latour in Pieces: An Intellectual Biography*. New York: Fordham University Press, 2014.

Schrödinger, Erwin. "The Present Situation in Quantum Mechanics: A Translation of Schrödinger's 'Cat Paradox' Paper." Trans. John D. Trimmer. *Proceedings of the American Philosophical Society* 124, no. 5 (1980): 323–338.

Schwartz, Heinrich. "Techno-Territories: The Spatial, Technological and Social Reorganization of Office Work." PhD diss., Massachusetts Institute of Technology, 2002.

Scott, Felicity D. "Securing Adjustable Climate." In *Climates: Architecture and the Planetary Imaginary*, ed. James Graham, Caitlin Blanchfield, Alissa Anderson, Jordan H. Carver, and Jacob Moore, 90–105. Zurich: Lars Müller, 2009.

Seamans, Robert C. Jr. Recorded interviews by Walter D. Sohier, Addison M. Rothrock, and Eugene M. Emme, on March 27, 1964, John F. Kennedy Library Oral History Program, Boston. https://archive1.jfklibrary.org/JFKOH/Seamans,%20Robert%20C.,%20Jr/JFKOH-RCS-01/JFKOH-RCS-01-TR.pdf.

Selections from the Norton Collection at RAND. Santa Monica, CA: RAND, 1991.

Sennett, Richard. *The Corrosion of Character: The Personal Consequences of Work in the New Capitalism*. London: W.W. Norton, 1998.

Serriano, Pierluigi. *The Creative Architect: Inside the Great Midcentury Personality Study*. New York: The Monacelli Press, 2016.

Shapin, Steven. *Never Pure: Historical Studies of Science as if It Was Produced by People with Bodies, Situated in Time, Space, Culture, and Society, and Struggling for Credibility and Authority*. Baltimore, MD: John Hopkins University Press, 2010.

Shapin, Steven. "Placing the View from Nowhere: Historical and Sociological Problems in the Location of Science." *Transactions of the Institute of British Geographers* 23, no. 1 (1998): 5–12.

Sheldon, Roy, and Egmont Arens. *Consumer Engineering: A New Technique for Prosperity*. New York: Arno Press, 1932.

Shulman, Julius, and Juergen Nogai. *Malibu: A Century of Living by the Sea*. New York: Harry Abrams Publishers, 2005.

Shinn, Terry. *L'École Polytechnique: 1794–1914*. Paris: Presses de la Fondation nationale des sciences politiques, 1980.

Simmel, George. *The Sociology of George Simmel*. Trans. K. Wolff. New York: The Free Press, 1950.

Simon, Herbert A. *Administrative Behavior: A Study of Decision-Making Processes in Administrative Organization*. New York: Macmillan Co., 1947.

Simon, Herbert A. *The Sciences of the Artificial*. Cambridge, MA: MIT Press, 1969.

Smith, Chris L. *Bare Architecture: A Schizoanalysis*. London: Bloomsbury, 2017.

Sobal, Jeffery, and Mary K. Nelson. "Commensal Eating Patterns: A Community Study." *Appetite* 41 (2003): 181–190.

Sontag, Susan. *Illness and Metaphor*. New York: Farrar, Straus and Giroux, 1978.

Stengers, Isabelle. *The Invention of Modern Science*. Trans. D. W. Smith. Minneapolis: The University of Minnesota Press, 2000.

Strugatskii, Arkadii, and Boris Strugatskii. *Ponedel'nik nachinaetsia v subboty: Skazka dlia nauchnykh rabotnikov mladwego vozrasta*. Moscow: Detskaia literatura, 1965.

Sturdy, Andrew, Mirela Schwarz, and Andre Spicer. "Guess Who's Coming to Dinner? Structures and Uses of Liminality in Strategic Management Consultancy." *Human Relations* 59, no. 7 (2006): 929–960.

Sulston, John, and Georgina Ferry. *The Common Thread: A Story of Science, Politics, Ethics and the Human Genome*. London: Corgi, 2003.

Tatarchenko, Ksenia. "Calculating a Showcase: Mikhail Lavrentiev, the Politics of Expertise, and the International Life of the Siberian Science-City." *Historical Studies in the Natural Sciences* 5 (2016): 592–632.

Tatarchenko, Ksenia. "'The Computer Does Not Believe in Tears:' Programming, Professionalization and the Gendering of Authority." *Kritika* 19, no. 4 (2017): 709–739.

Taylor, David. *The Right Formula*. Manchester: Manchester University Press, 2016.

Taylor, William M. *The Vital Landscape: Nature and the Built Environment in Nineteenth-Century Britain*. Aldershot: Ashgate Publishing, 2004.

Tcharfas, Julia. "Science of Rehearsal." In *Cosmonauts: Birth of the Space Age*, ed. Doug Millard. London: Scala Arts Publishers, Inc, 2015.

Thelen, Tatjana. "Lunch in an East German Enterprise—Differences in Eating Habits as Symbols of Collective Identities." *Zeitschrift fur Ethnologie* 131 (2006): 51–70.

Thorp, Holden, and Buck Goldstein. *Engines of Innovation: The Entrepreneurial University in the Twenty-First Century*. Chapel Hill: The University of North Carolina Press, 2010.

Tromly, Benjamin. *Making the Soviet Intelligentsia: Universities and Intellectual Life under Stalin and Khrushchev*. Cambridge: Cambridge University Press, 2013.

Tucker, George. *A Voyage to the Moon: With Some Account of the Manners and Customs, Science and Philosophy, of the People of Morosofia, and Other Lunarians*. New York: Elam Bliss, 1827. Project Gutenberg, http://www.gutenberg.org/cache/epub/10005/pg10005-images.html.

Turner, Christopher. "Ingesting the Biosphere." *Cabinet Magazine*, no. 41, 2011. http://cabinet magazine.org/issues/41/turner.php.

Turner, Fred. *From Counterculture to Cyberculture: Stewart Brand, the Whole Earth Network and the Rise of Digital Utopianism*. Chicago: University of Chicago Press, 2006.

Turner, Paul Vernable. *Campus: An American Planning Tradition*. Cambridge, MA: MIT Press, 1984.

United Nations. "World Urbanization Prospects: The 2011 Revision." *Department of Economic and Social Affairs Population Division*, 2012. http://www.un.org/en/development/desa/population/ publications/pdf/urbanization/WUP2011_Report.pdf.

Vasilyeva, Zinaida. "Der unauffällige Staat: Die Infrastruktur der Amateurverbände in der Zeit des Spätsozialismus." In *Hochkultur für das Volk? Literatur, Kunst und Musik in der Sowjetunion aus kulturgeschichtlicher Perspektive*, ed. Igor Narskij, 213–234. Berlin: De Gruyter/Oldenbourg, 2018.

Veblen, Thorstein. *The Theory of the Leisure Class: An Economic Study in the Evolution of Institutions*. New York: Macmillan, 1899.

Venter, Craig. *A Life Decoded: My Genome, My Life*. New York: Viking, 2007.

Veysey, Laurence R. *The Communal Experience: Anarchist and Mystical Communities in Twentieth Century America*. Chicago: University of Chicago Press, 1978.

Von Geldern, James. *Bolshevik Festivals, 1917–1920*. Berkeley: University of California Press, 1993.

Walford, Roy L. "Biosphere 2 as Voyage of Discovery: The Serendipity from Inside." *Bioscience* 52, no. 3 (March 2002): 259–263.

Waring, Amanda. "Health Club Use and 'Lifestyle': Exploring the Boundaries between Work and Leisure." *Leisure Studies* 27 (2008): 295–309.

Watson, Elizabeth L. *Grounds for Knowledge: A Guide to Cold Spring Harbor Laboratory's Landscapes and Buildings.* Cold Spring Harbor, NY: Cold Spring Harbor Laboratory Press, 2008.

Watson, Elizabeth L. *Houses for Science: A Pictorial History of Cold Spring Harbor Laboratory.* Cold Spring Harbor, NY: Cold Spring Harbor Laboratory Press, 1991.

Watson, James. *Avoid Boring People: Lessons from a Life in Science.* Oxford: Oxford University Press, 2007.

Watson, Traci. "Can Basic Research Ever Find a Good Home in Biosphere 2?," *Science* 259, no. 5102 (March 1993): 1688–1689.

Weber, Max. *The Protestant Ethic and the Spirit of Capitalism.* Trans. T. Parsons. New York: Charles Scribner's Sons, [1905] 1963.

Weinberg, Alvin M. "Impact of Large-Scale Science on the United States." *Science* 134 (1961): 161–164.

Welsch, Wolfgang. *Undoing Aesthetics.* London: Sage, 1997.

Wen Wong, Kai, and Tim Bunnell. "'New Economy' Discourse and Spaces in Singapore: A Case Study of One-North." *Environment & Planning A* 38 (2006): 38–78.

Wilson, E. O. *Naturalist.* Washington, DC: Island Press, 1994.

Wooldridge, E. T. *Winged Wonders: The Story of the Flying Wings.* Washington, DC: Smithsonian Institution Press, 1983.

Wynne, Derek. *Leisure, Lifestyle and the New Middle Class: A Case Study.* London: Routledge, 1998.

Yaneva, Albena. "A Building Is a Multiverse." In *Making Things Public*, ed. Bruno Latour and Peter Weibel, 530–535. Cambridge, MA: MIT Press, 2005.

Yaneva, Albena. "Challenging the Visitor to Get the Image." In *Iconoclash, Beyond the Image War in Science, Religion and Art*, ed. Bruno Latour and Peter Weibel, 421–422. Cambridge, MA: MIT Press, 2002.

Yaneva, Albena. "Is the Atrium More Important than the Lab? Designer Buildings for New Cultures of Creativity. In *Geographies of Science*, ed. Peter Meusburger, David N. Livingstone, and Heike Jöns, 139–151. Dordrecht: Springer, 2010.

Yaneva, Albena. *The Making of a Building: A Pragmatist Approach to Architecture.* Oxford: Peter Lang Publishers, 2009.

Yaneva, Albena. "Scaling Up and Down: Extraction Trials in Architectural Design." *Social Studies of Science* 35, no. 6 (2005): 867–894.

Young, Stephanie. "'Would Your Questions Spoil My Answers?': Arts and Technology at the RAND Corporation." In *Where Minds and Matters Meet: Technology and California and the West*, ed. Volker Janssen, 293–320. Berkeley: University of California Press, 2012.

Yurchak, Aleksei. *Everything Was Forever, Until: The Last Soviet Generation*. Princeton, NJ: Princeton University Press, 2006.

Zaera-Polo, Alejandro. "The Politics of the Envelope: A Political Critique of Materialism." *ArchiNed* 17 (2008): 76–105.

Zimmerman, Michael. "Review: Biosphere 2: Long on Hype, Short on Science." *Ecology* 73, no. 2 (April 1992): 713–714.

Zubok, Vladislav. *Zhivago's Children: The Last Russian Intelligentsia*. Cambridge, MA: Belknap Press, 2009.

Contributors

Kathleen Brandt, Syracuse University

Kathleen Brandt is design principal at KBL Studio with Brian Lonsway and research assistant professor in the School of Education at Syracuse University. With a background in sculpture, information arts, and electronic media installation, Brandt works between the fields of design, art, and education to develop and promote design systems that respond to current issues of social and cultural change.

Russell Hughes, University of Queensland

Russell Hughes is Japan Society for the Promotion of Science Visiting Professor in the Institute of Oriental and Occidental Cultures at Kansai University, and Honorary Fellow in the School of Architecture at the University of Queensland. With a background in art, architecture, philosophy, and science, his current research examines digital urbanism and its impact on biopolitics, in particular population health.

Tim Ivison, SCI-Arch

Tim Ivison is an interdisciplinary scholar working on the political ecology of modern cities and urban planning. He teaches in both Liberal Arts and History + Theory at SCI-Arc, and in the department of Humanities and Sciences at Art Center College of Design. Recent collaborations with the artist Julia Tcharfas include *T.E.O.T.W.A.W.K.I.* and *Operations Theater* at Before Present, Los Angeles. Ivison holds a BA in visual and critical studies and a BFA in studio practice from the School of the Art Institute of Chicago. In 2016 he completed a PhD in humanities and cultural studies at the London Consortium, with a thesis on biopolitics and environmental discourse in modern British town planning. From 2014 to 2016 Ivison was an Andrew W. Mellon Foundation Researcher at the Canadian Centre for Architecture, Montreal.

Sandra Kaji-O'Grady, University of Queensland

Sandra Kaji-O'Grady is dean and head of the School of Architecture at the University of Queensland. Kaji-O'Grady's research on the transfer of ideas and techniques between art, architecture, and the experimental sciences has been published in leading journals including the *Journal of Architecture, Journal of Architectural Education, Architecture Research Quarterly, Architecture and Culture,* and *Le Journal Spéciale'Z.*

Stuart W. Leslie, Johns Hopkins University

Stuart W. Leslie is professor of history of science at Johns Hopkins University. His books include *The Cold War and American Science: The Military-Industrial-Academic Complex at MIT and Stanford* (Columbia University Press, 1994) and *Boss Kettering: Wizard of General Motors* (Columbia University Press, 1986). His research focuses primarily on the history of modern American science and technology, in recent years concentrating on the architecture of corporate headquarters, research laboratories, and healthcare facilities. He also studies observatories, especially public astronomy, with attention to matters of design.

Brian Lonsway, Syracuse University

Brian Lonsway is associate professor of architecture at Syracuse University and a partner with Kathleen Brandt in KBL Studios, an experimental media environment and situation room for complex thinking designed to explore and foster alternative practices of transdisciplinary design inquiry. His current research is centered on themed landscapes, the inherent transdisciplinarity of design, and the intersection of disciplinary and professional identities with alternative models of design practice. Lonsway is the author of *Making Leisure Work: Architecture and the Experience Economy* (Routledge, 2009) and has published in *Convergence* and in the *Journal of Architectural Education.*

Sean O'Halloran, University of Western Australia

Sean O'Halloran is an adjunct senior lecturer in pathology and laboratory medicine at the University of Western Australia, researching in the field of pharmacology, and drug transfer in breast milk. He is also senior clinical specialist in clinical pharmacology and toxicology at the Sir Charles Gairdner Hospital.

Simon Sadler, University of California, Davis

Simon Sadler is professor of design at the University of California, Davis. He has been a UC Davis Chancellor's Fellow, a fellow of the University of California Humanities Research Institute, and a fellow of the Paul Mellon Centre for Studies in British Art,

London. He has served on the editorial boards of the *Journal of Architectural Education* and *The Architect's Newspaper*. His books include *Archigram: Architecture without Architecture* (MIT Press, 2005), *Non-Plan: Essays on Freedom, Participation and Change in Modern Architecture and Urbanism* (Architectural Press, 2000, coeditor Jonathan Hughes), and *The Situationist City* (MIT Press, 1998).

Chris L. Smith, University of Sydney

Chris L. Smith is associate professor in architectural design and technê in the School of Architecture at the University of Sydney. His research concerns the interdisciplinary nexus of philosophy, biology, and architectural theory. Smith has published on the political philosophy of Deleuze and Guattari; technologies of the body; and the influence of "the eclipse of Darwinism" phase on contemporary architectural theory. He is the author of *Architecture in the Space of Flows* (Routledge, 2013, coeditor with Andrew Ballantyne) and *Bare Architecture: A Schizoanalysis* (Bloomsbury, 2017).

Nicole Sully, University of Queensland

Nicole Sully is a senior lecturer in the School of Architecture at the University of Queensland. Her research focuses on the relationship of architecture and memory; pathologies of place; heritage; modern architecture and the history of Australian art and architecture. Her work has been published in *The Architect*, *Fabrications*, and *ARQ: Architectural Research Quarterly*. Sully's books include *Shifting Views: Selected Essays on the Architectural History of Australia and New Zealand* (University of Queensland Press, 2008, coedited with Andrew Leach and Antony Moulis), and *Out of Place (Gwalia): Occasional Essays on Australian Regional Communities and Built Environments in Transition* (University of Western Australia Press, 2014, coedited with Philip Goldswain and William M. Taylor).

Ksenia Tatarchenko, Geneva University

Ksenia Tatarchenko is a lecturer at the Institute of Global Studies, Geneva University. Her research is on the transnational history of computing; science in the Russian and Soviet Empires, in particular Siberia; urban history; gender; and the Cold War. She received her PhD from the History of Science Program, History Department, Princeton University. Tatarchenko was a postdoctoral fellow at the Harriman Institute, Columbia University, and a visiting assistant professor of history at NYU Shanghai.

William Taylor, University of Western Australia

William Taylor is professor in the UWA School of Design and a registered architect. His research is in the history and theories of the built environment, buildings in social and political contexts, philosophy and architecture, ships and the sea, and disaster. Taylor's books include *Prospects for an Ethics of Architecture* (London: Routledge, 2011, coedited with Michael Levine), *The Vital Landscape: Nature and the Built Environment in Nineteenth-Century Britain* (Ashgate Press, 2004), and *The "Katrina Effect": On the Nature of Catastrophe* (Bloomsbury 2015, coedited with Michael P. Levine, Oenone Rooksby, and Joely-Kym Sobott).

Julia Tcharfas

Julia Tcharfas is an artist and curator based in Los Angeles, researching the design and performance of artificial natures. Recent projects include *Before Present* at Operations Theater, Pasadena (2017); *Cosmonauts: Birth of the Space Age*, Science Museum, London (2015); *Interspecies Communication*, Serpentine Gallery, London (2015); and *Systems Thinking from the Inside*, Chisenhale Gallery (2013). Her books include the exhibition catalogue *Cosmonauts: Birth of the Space Age* (Scala Press, 2015) and her work has been published in *Flash Art*, *Science Museum Group Journal*, *Living in the Future*, *Monaco* magazine, and *Waterfall*.

Albena Yaneva, University of Manchester

Albena Yaneva is professor of architectural theory and director of the Manchester Architecture Research Group (MARG) at the Urban Institute, University of Manchester. Her research is intrinsically transdisciplinary and crosses the boundaries of science studies, cognitive anthropology, architectural theory, and political philosophy. She is the author of several books: *The Making of a Building* (Peter Lang, 2009), *Made by the OMA: An Ethnography of Design* (010 Publishers, 2009), *Mapping Controversies in Architecture* (Ashgate, 2012), *Five Ways to Make Architecture Political: An Introduction to the Politics of Design Practice* (Bloomsbury, 2017), and editor with Alejandro Zaera-Polo of *What Is Cosmopolitical Design?* (Routledge, 2015). Yaneva is the recipient of the RIBA President's Award for Research (2010).

Stelios Zavos, University of Manchester

Stelios Zavos is a PhD candidate in architecture and a teaching assistant at the University of Manchester. His research traverses the boundaries of architectural studies and anthropology, focusing on the material semiotics of inhabitation in urban residential buildings. He earned his master's degree in urban strategies and design from the University of Edinburgh in 2015. Prior to that, Zavos trained as a civil engineer with specialization in spatial planning and practiced as an engineering consultant for infrastructure and residential projects in Greece.

Index